AF817634

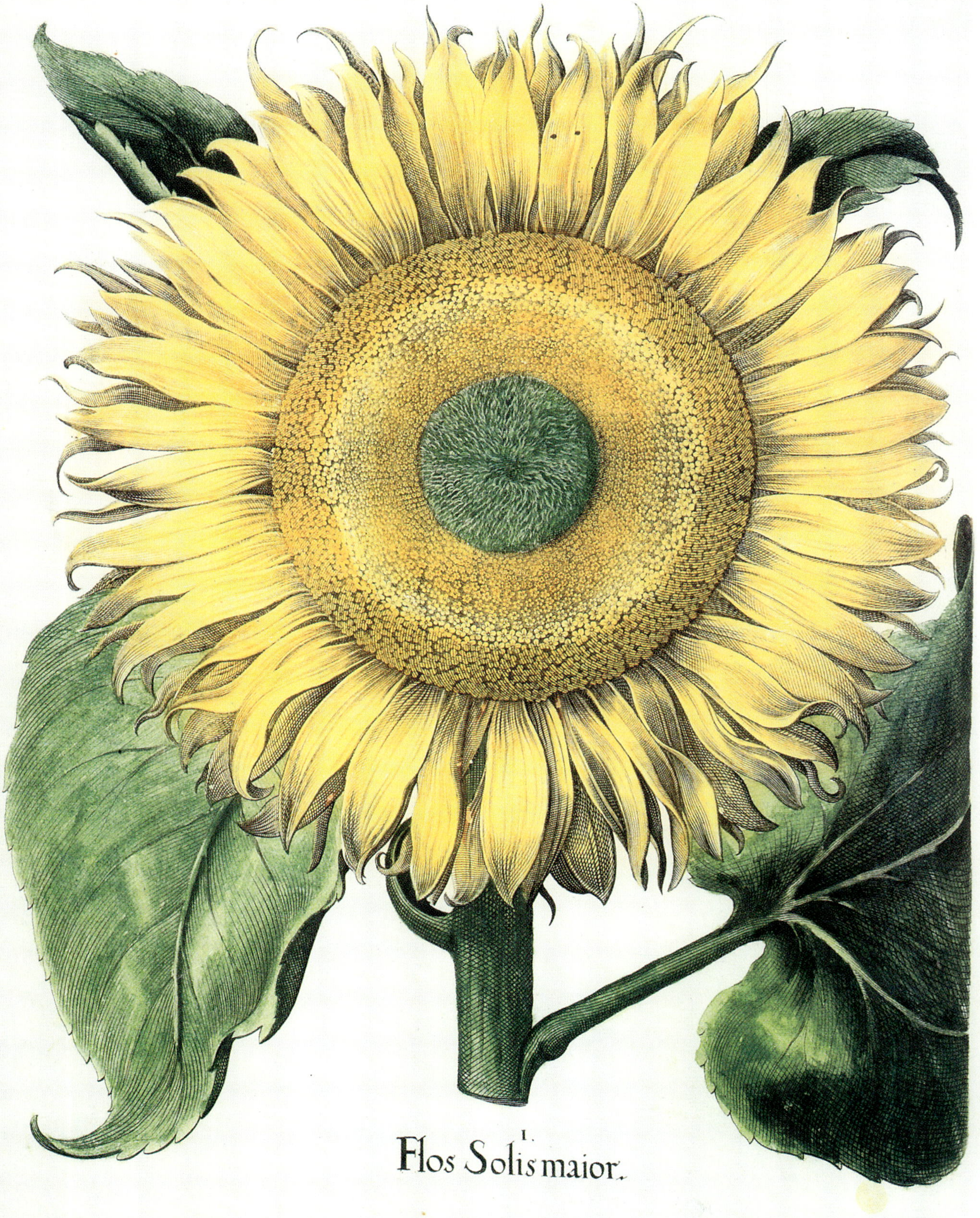

Common Sunflower
(Helianthus annuus)

Besler's Book of Flowers & Plants

73 Full-Color Plates from *Hortus Eystettensis*, 1613

Basilius Besler

DOVER PUBLICATIONS, INC.
Mineola, New York

Publisher's Note

Originally published in 1613 by apothecary Basilius Besler, the three-volume *Hortus Eystettensis* was a spectacular compilation of engravings depicting thousands of exotic plants from the garden at Eichstätt in Germany, created by the Prince Bishop Johann Konrad von Gemmingen. The large-folio copperplate engravings were the result of Besler's monumental undertaking—to create a visual record of what was likely the first comprehensive botanical garden devoted to flowering plants. With the assistance of as many as ten engravers, Besler documented this vast and impressive garden, illustrating each plant as it sprouted and bloomed throughout the four seasons.

The present volume reproduces, in full color, a selection of seventy-three plates from Besler's grand work. The magnificent engravings that appear here represent a wide variety of flowers and plants found in the Prince Bishop's lush garden at Eichstätt. This new compilation of exquisite illustrations, carefully selected from one of the most treasured natural history editions of all time, will enthrall anyone who appreciates the glorious beauty of floral art and botanicals.

Copyright

Copyright © 2007 by Dover Publications, Inc.
All rights reserved.

Bibliographical Note

This Dover edition, first published in 2007, is a new selection of 73 plates from *Hortus Eystettensis,* originally published as a three-volume set in Nuremberg in 1613. The Publisher's Note has been specially written for the present edition.

DOVER *Pictorial Archive* SERIES

This book belongs to the Dover Pictorial Archive Series. You may use the designs and illustrations for graphics and crafts applications, free and without special permission, provided that you include no more than four in the same publication or project. (For permission for additional use, please write to Permissions Department, Dover Publications, Inc., 31 East 2nd Street, Mineola, N.Y. 11501.)

However, republication or reproduction of any illustration by any other graphic service, whether it be in a book or in any other design resource, is strictly prohibited.

Library of Congress Cataloging-in-Publication Data

Besler, Basilius, 1561–1629.
[Hortus Eystettensis. Plates. Selections]
Besler's book of flowers and plants : 73 full-color plates from "Hortus Eystettensis," 1613 / Basilius Besler.
p. cm. — (Dover pictorial archive series)
"This Dover edition, first published in 2007, is a new selection of 73 plates from Hortus Eystettensis, originally published as a three-volume set in Nuremberg in 1613. The Publisher's Note has been specially written for the present edition."
ISBN-13: 978-0-486-46005-5
ISBN-10: 0-486-46005-3
1. Besler, Basilius, 1561–1629. Hortus Eystettensis—Illustrations. 2. Hortus Eystettensis (Germany)—Early works to 1800. 3. Botanical illustration—Early works to 1800. 4. Botany—Pre-Linnean works. I. Title.

QK41.B52 2007
580.22'2—dc22

2007010628

Manufactured in the United States of America
Dover Publications, Inc., 31 East 2nd Street, Mineola, N.Y. 11501

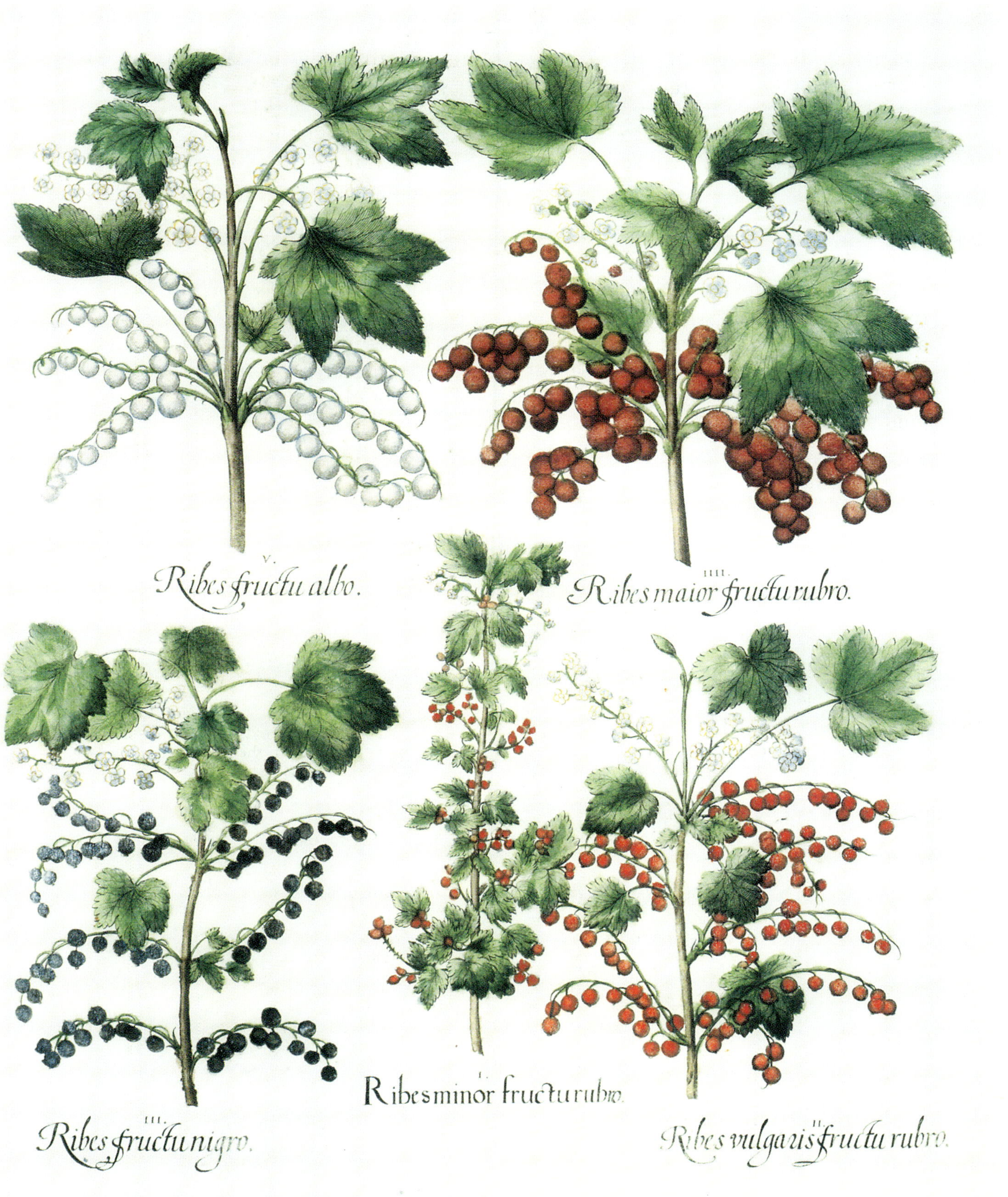

I. Alpine currant; Mountain currant *(Ribes alpinum)*. **II, IV (IIII), V.** Red currant *(Ribes rubrum)*. **III.** Black currant *(Ribes nigrum)*.

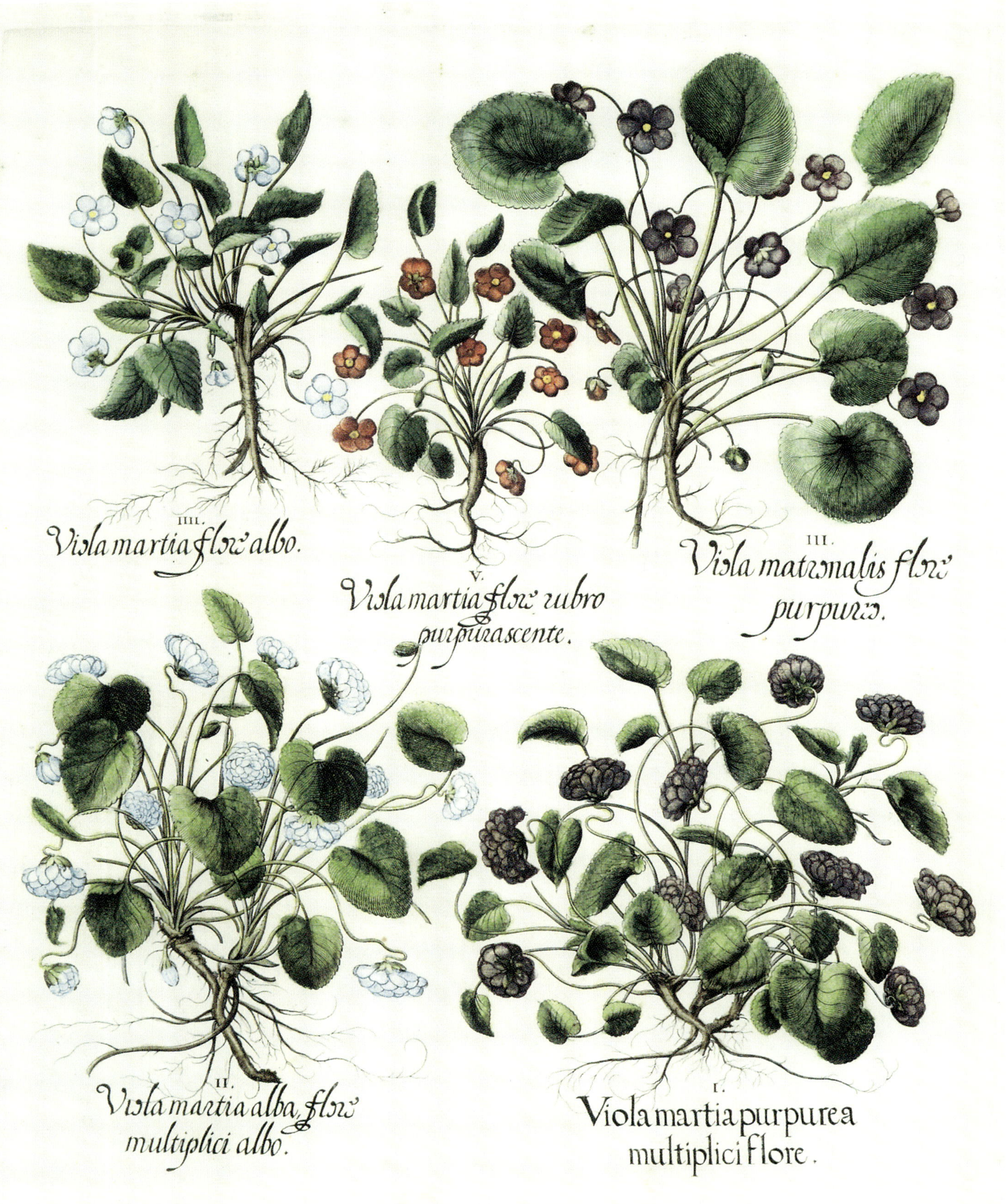

I–V. English violet; Sweet violet
(Viola odorata).

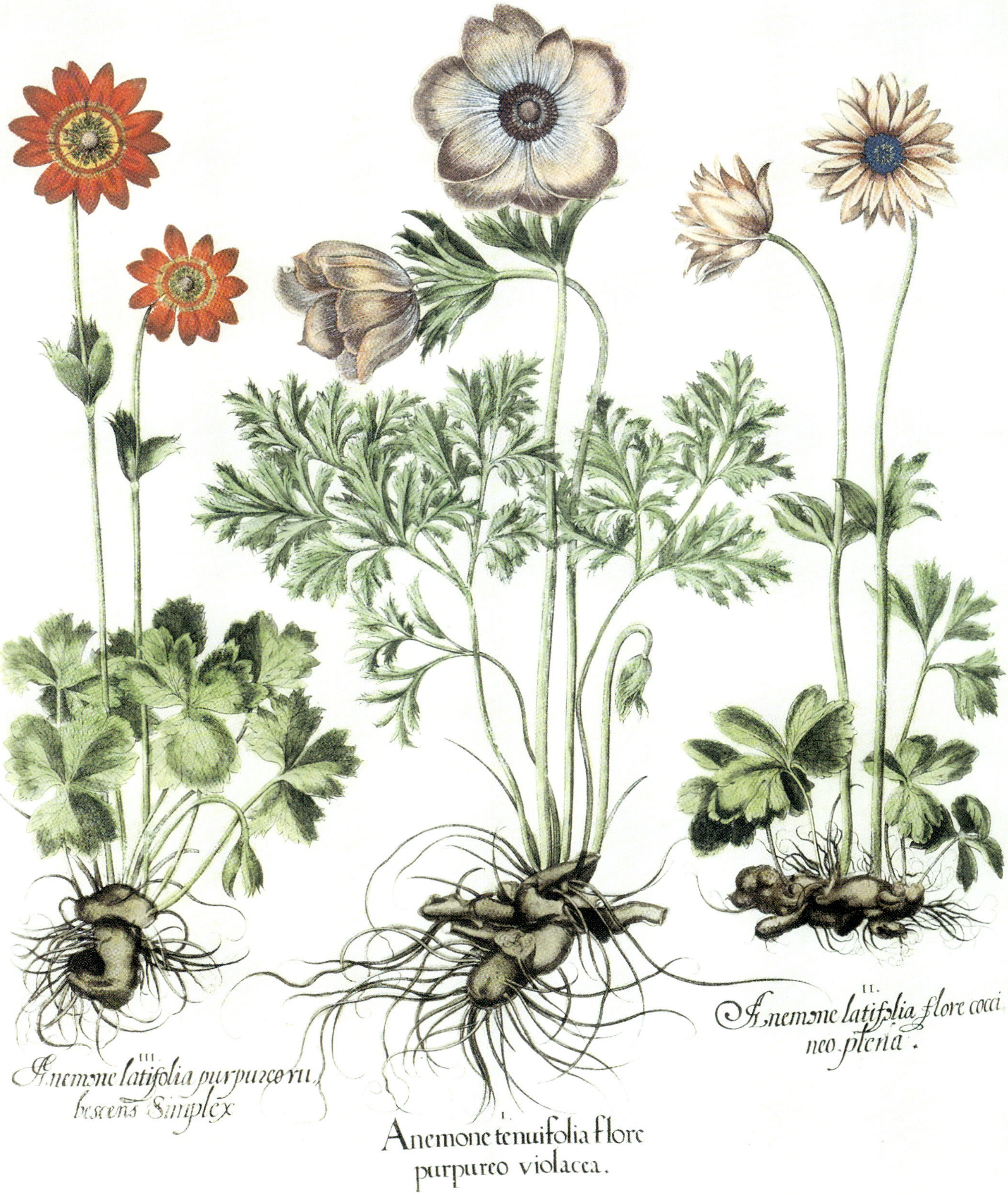

I. Windflower *(Anemone coronaria)*. **II, III.** Windflower *(Anemone hortensis)*.

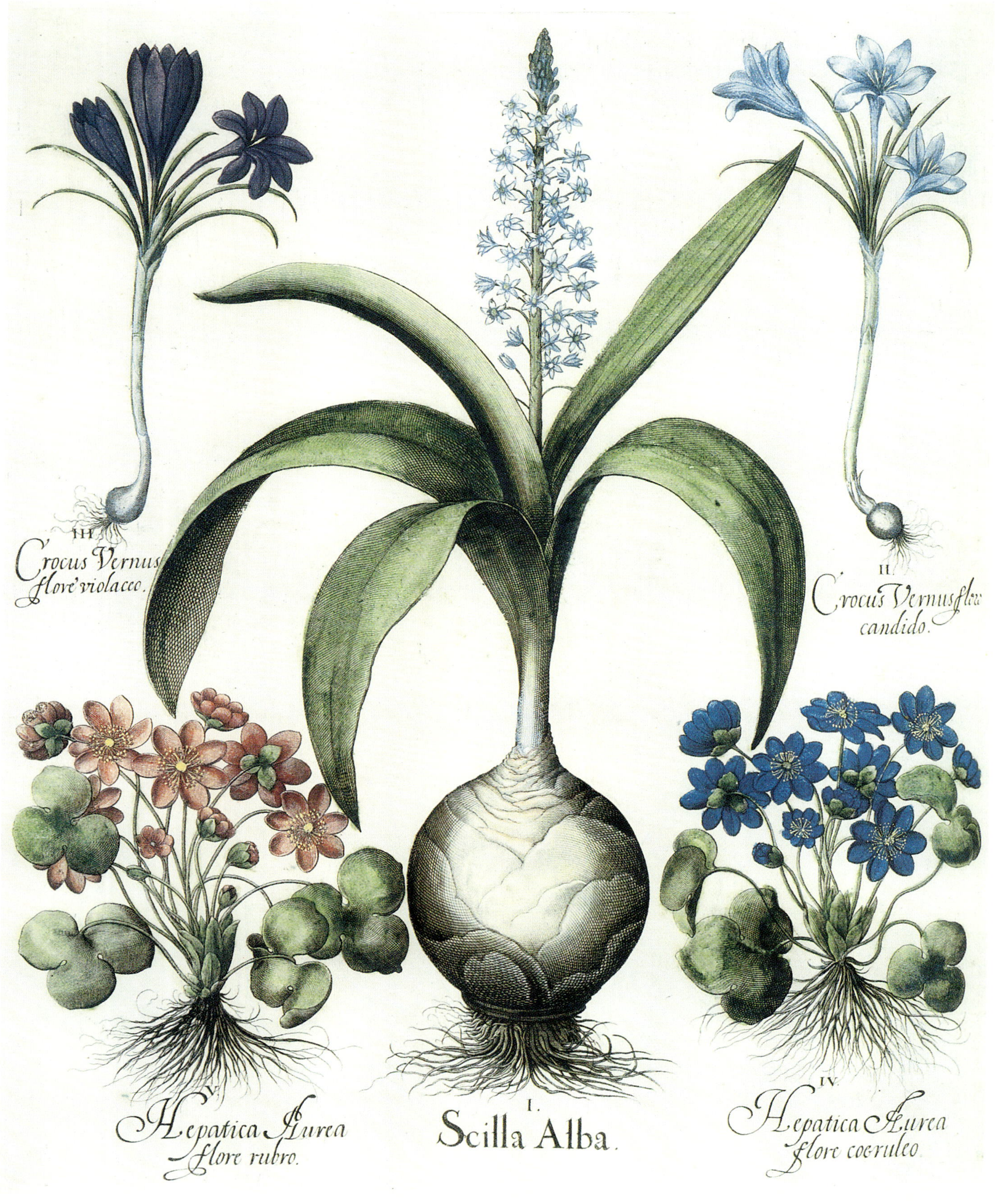

I. Sea onion *(Urginea maritima).* **II, III.** Crocus *(Crocus vernus).*
IV, V. Blue anemone *(Hepatica nobilis).*

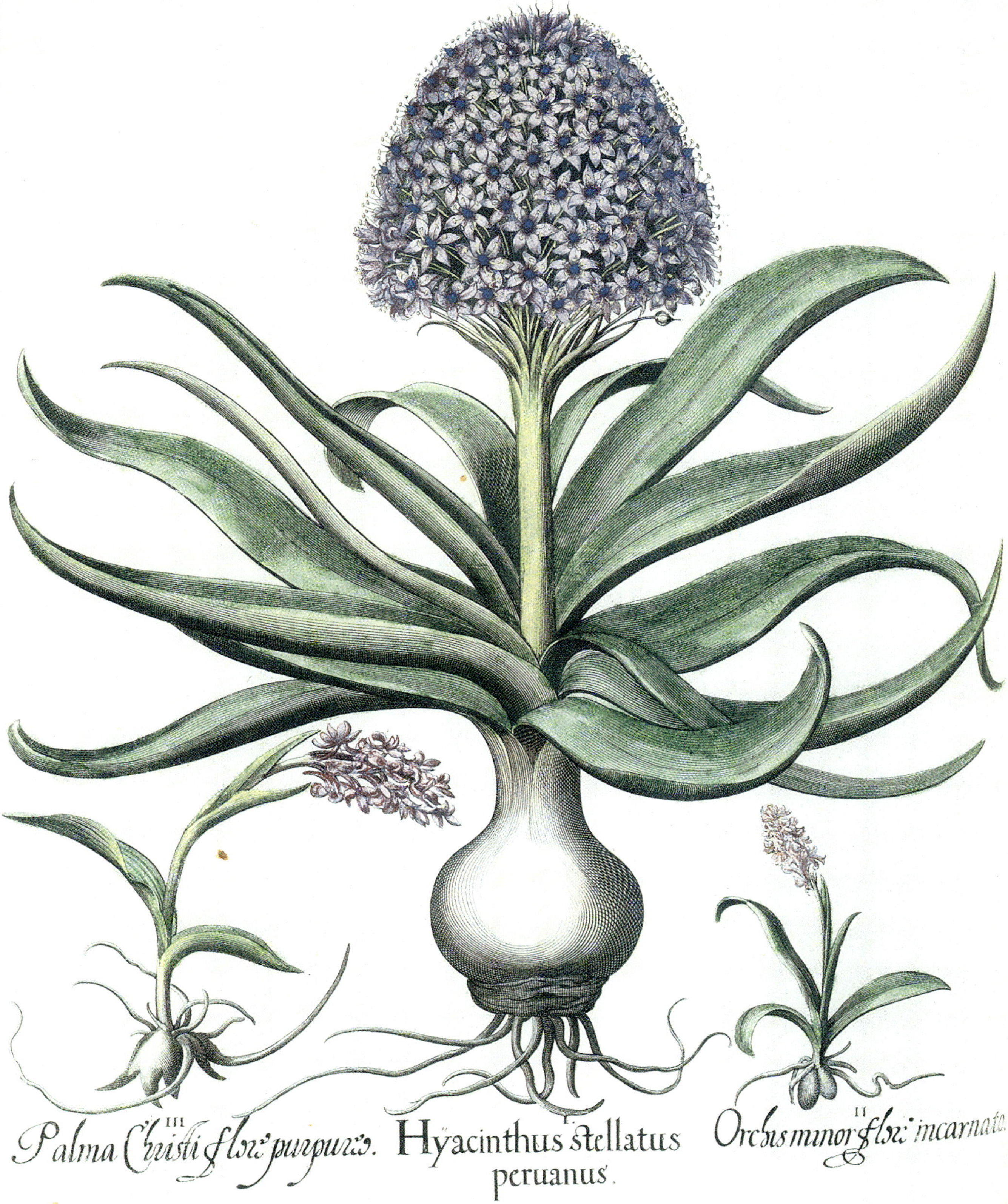

I. Squill of Peru *(Scilla peruviana).* **II.** Toothed orchid *(Orchis tridentata).*
III. Broad-leaved marsh orchid *(Dactylorhiza majalis).*

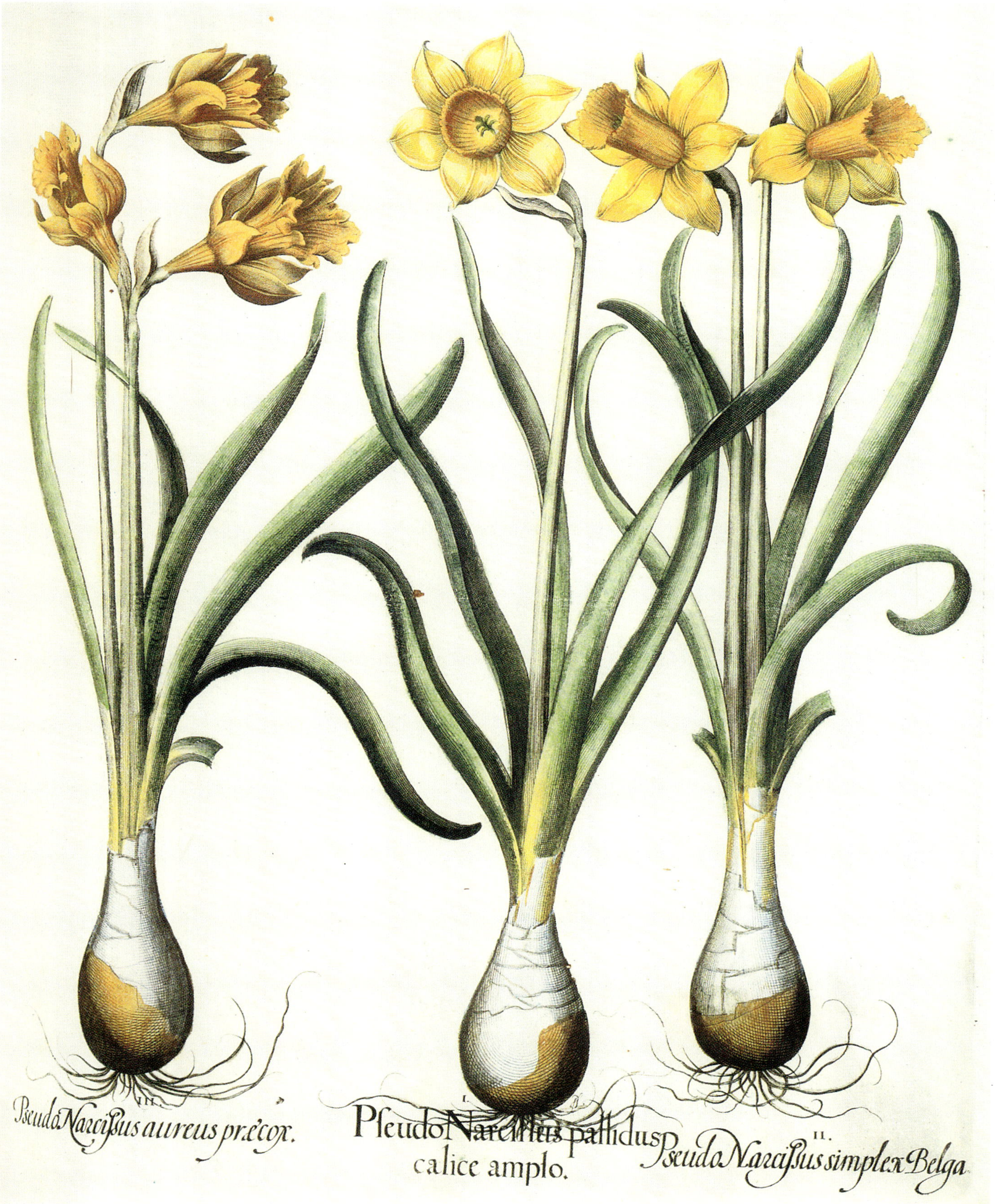

I, II. Daffodil *(Narcissus* x *incomparabilis).* **III.** Wild daffodil *(Narcissus pseudonarcissus).*

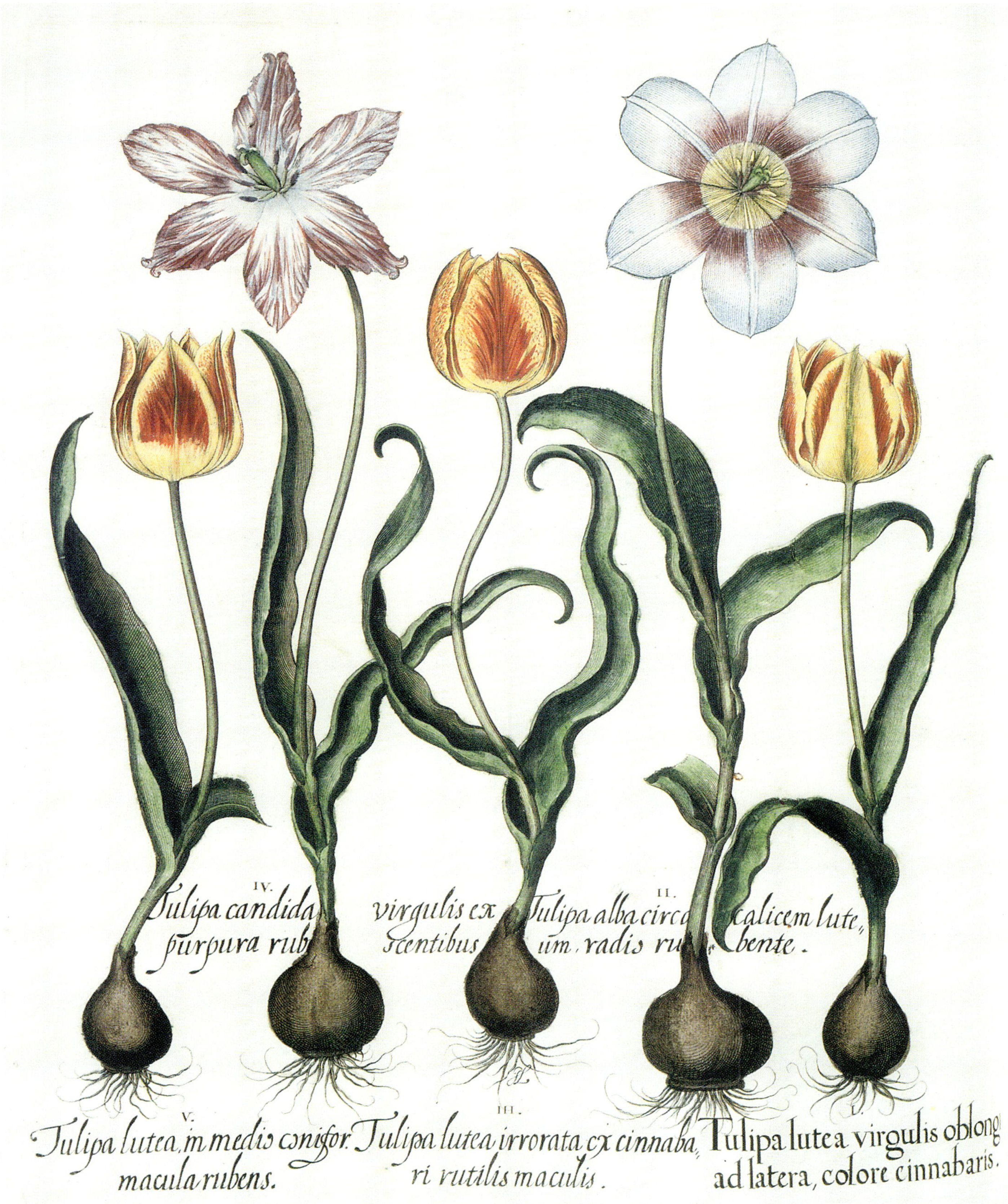

I. Yellow tulip, flamed red and green *(Tulipa spec.)*. **II.** White tulip, yellow and carmine inside *(Tulipa spec.)*. **III.** Golden and brick red variegated tulip *(Tulipa spec.)*. **IV.** Red and white striped tulip *(Tulipa spec.)*. **V.** Golden yellow tulip with purple sheen *(Tulipa spec.)*.

Plate 8

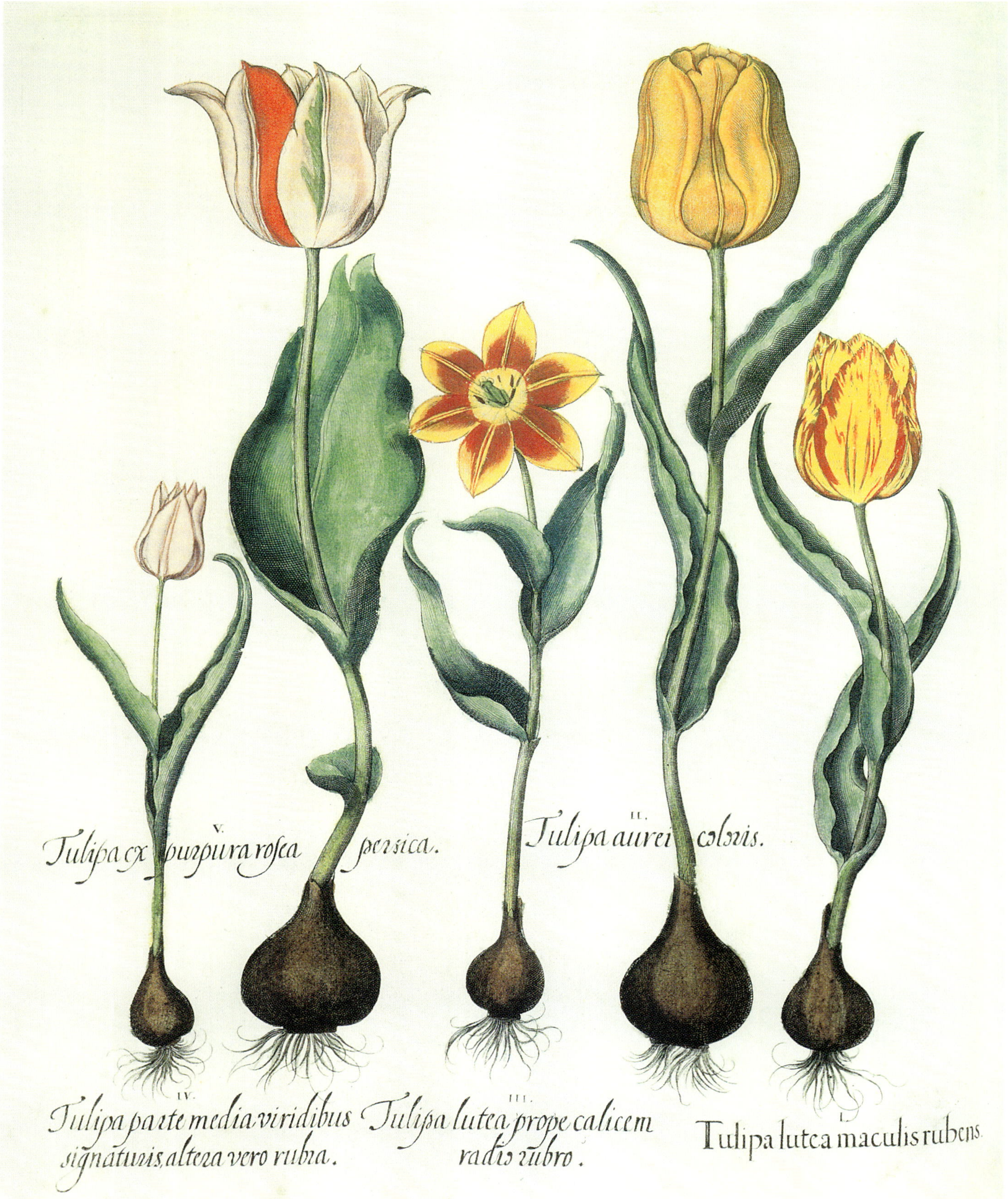

I. Early yellow and red flamed tulip *(Tulipa spec.)*. **II.** Early yellow tulip with large orange-colored flowers *(Tulipa spec.)*. **III.** Early yellow tulip with brick red rim *(Tulipa spec.)*. **IV.** Late red tulip, bursting into flower *(Tulipa spec.)*. **V.** Early, single, pale pink tulip *(Tulipa spec.)*.

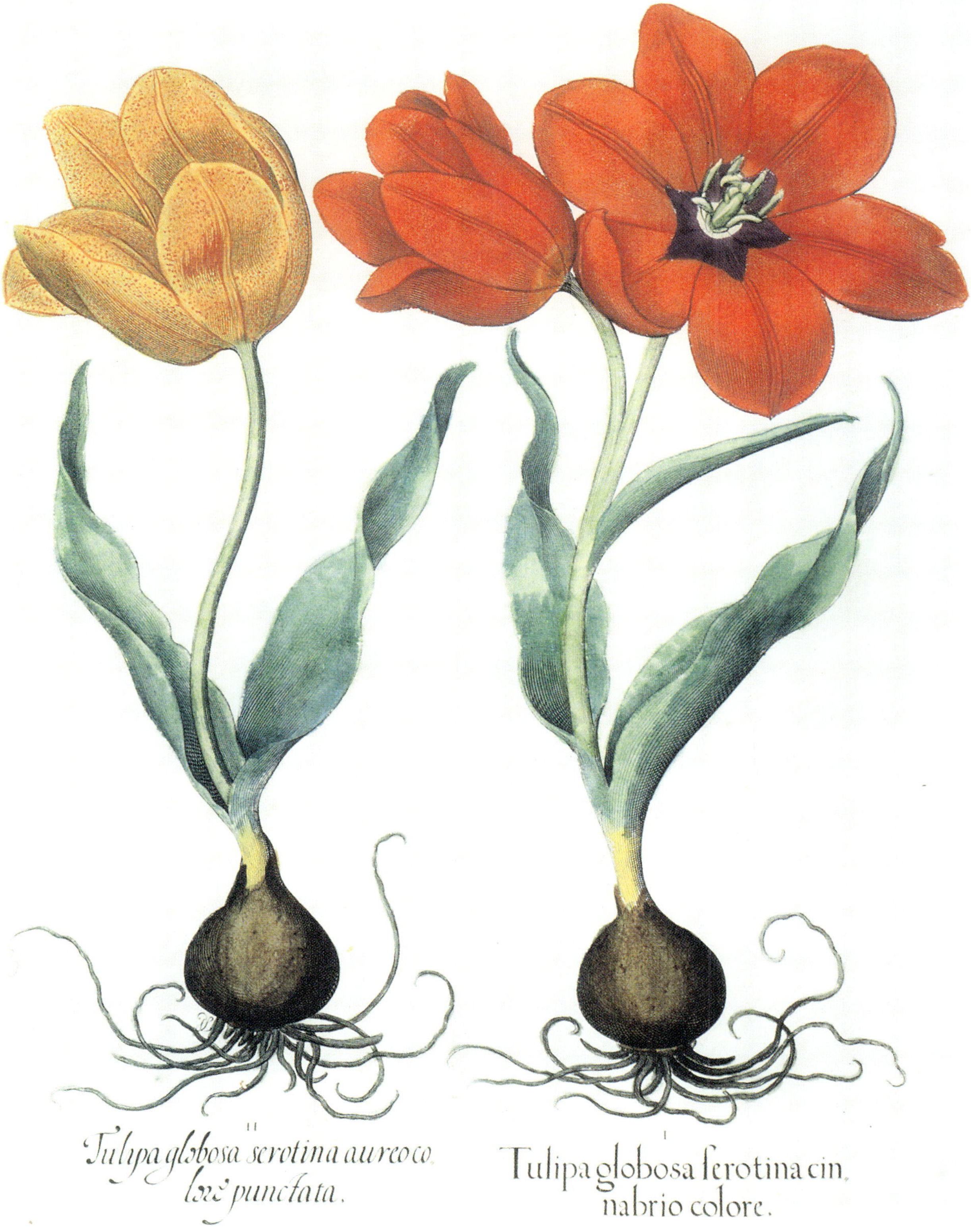

I. Late tulip with large red flower *(Tulipa spec.)*. **II.** Tulip with yellow-feathered petals *(Tulipa spec.)*.

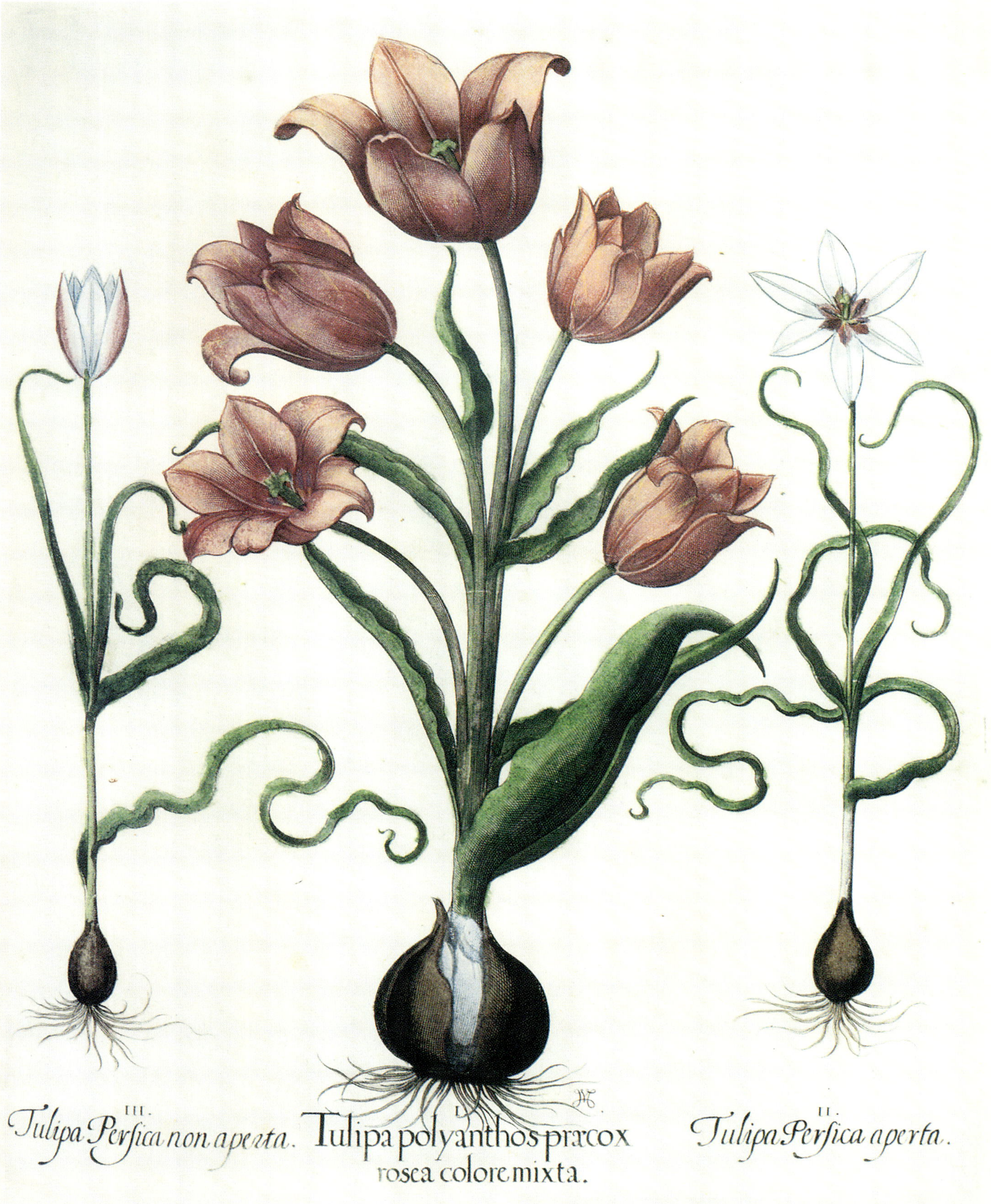

I. Multi-flowered red tulip *(Tulipa spec.)*. **II.** White botanical tulip, flower open *(Tulipa spec.)*. **III.** White botanical tulip, flower closed *(Tulipa spec.)*.

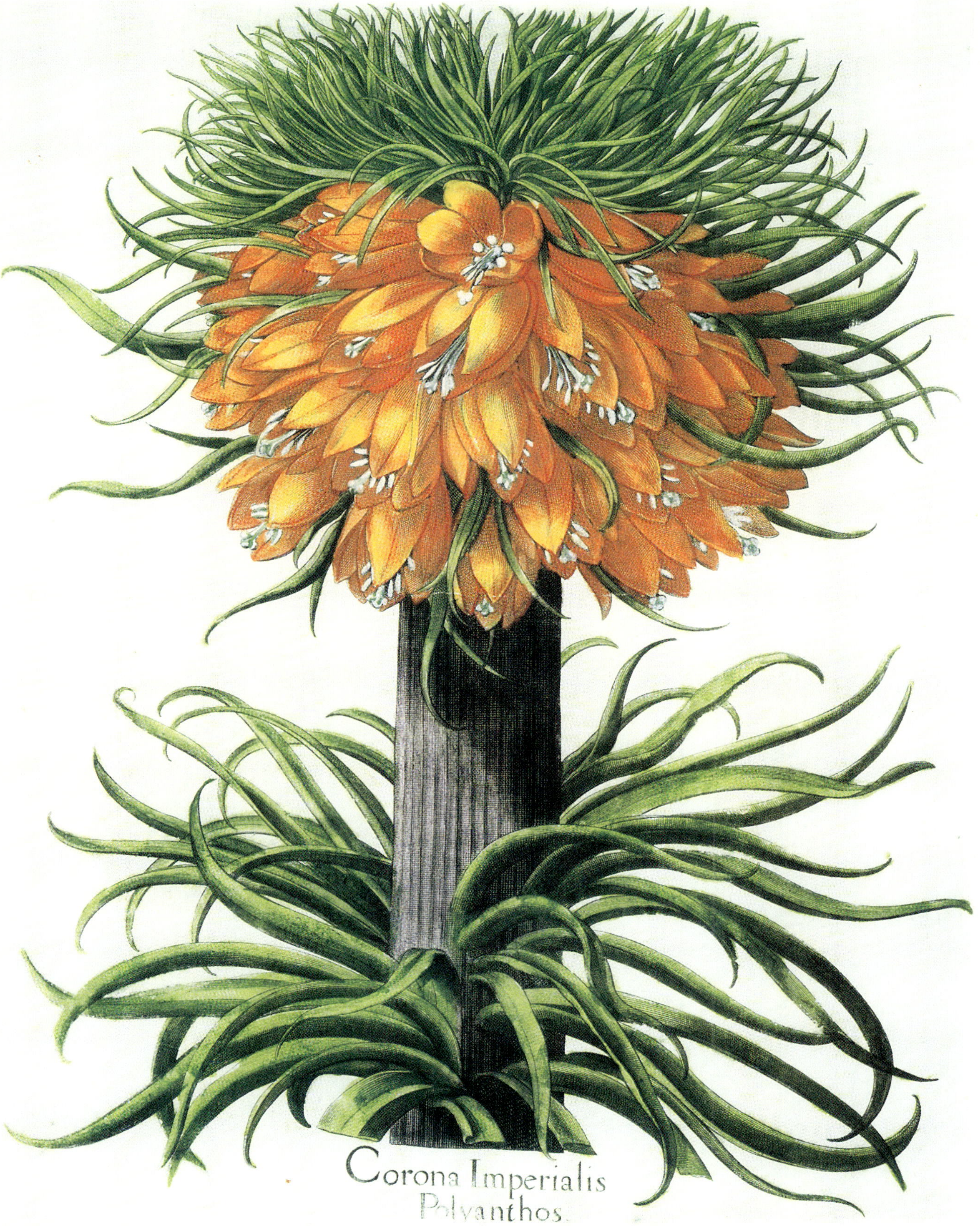

Multi-flowered crown imperial
(Fritillaria imperialis)

Plate 12

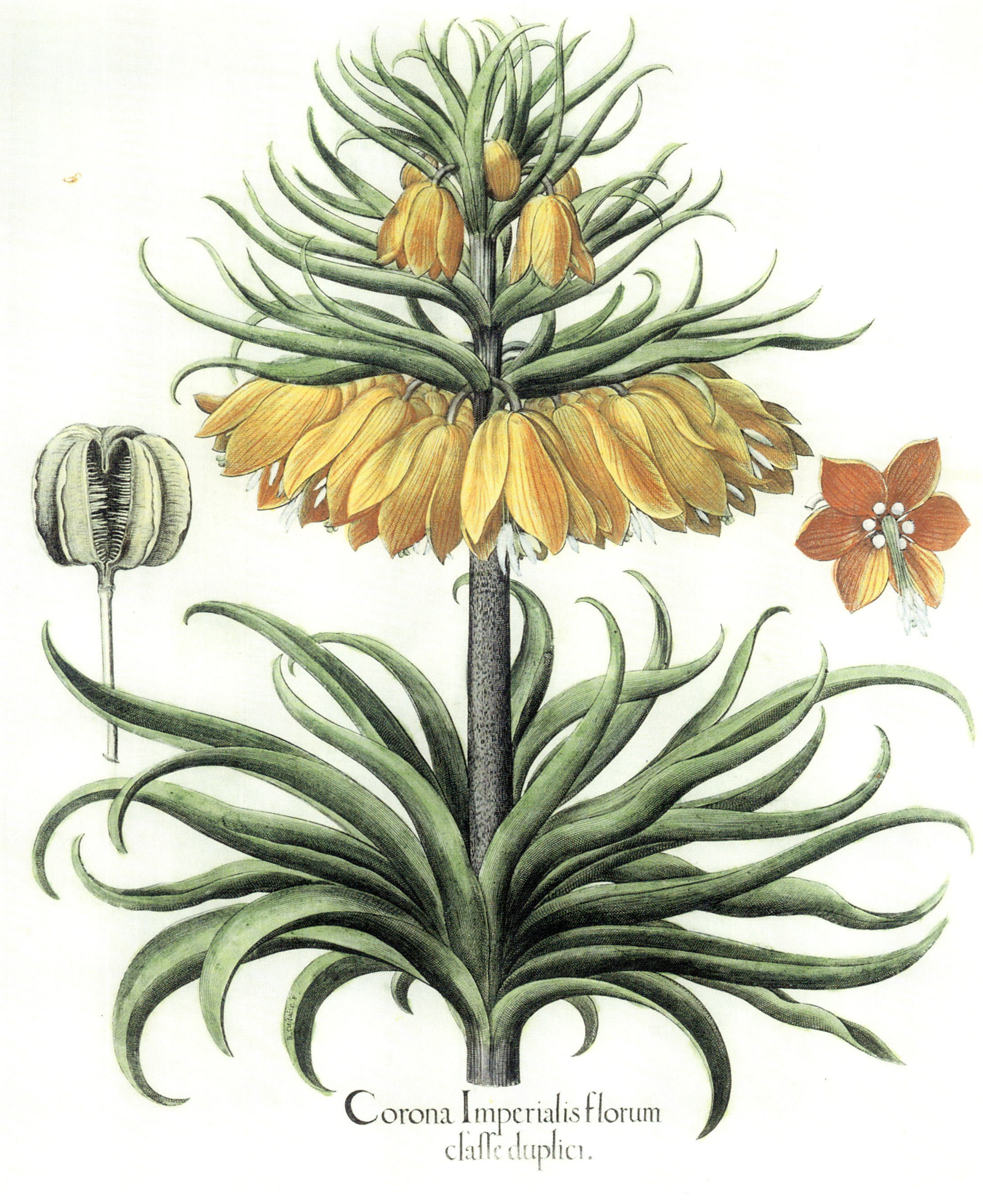

Crown imperial with two crowns
(Fritillaria imperialis)

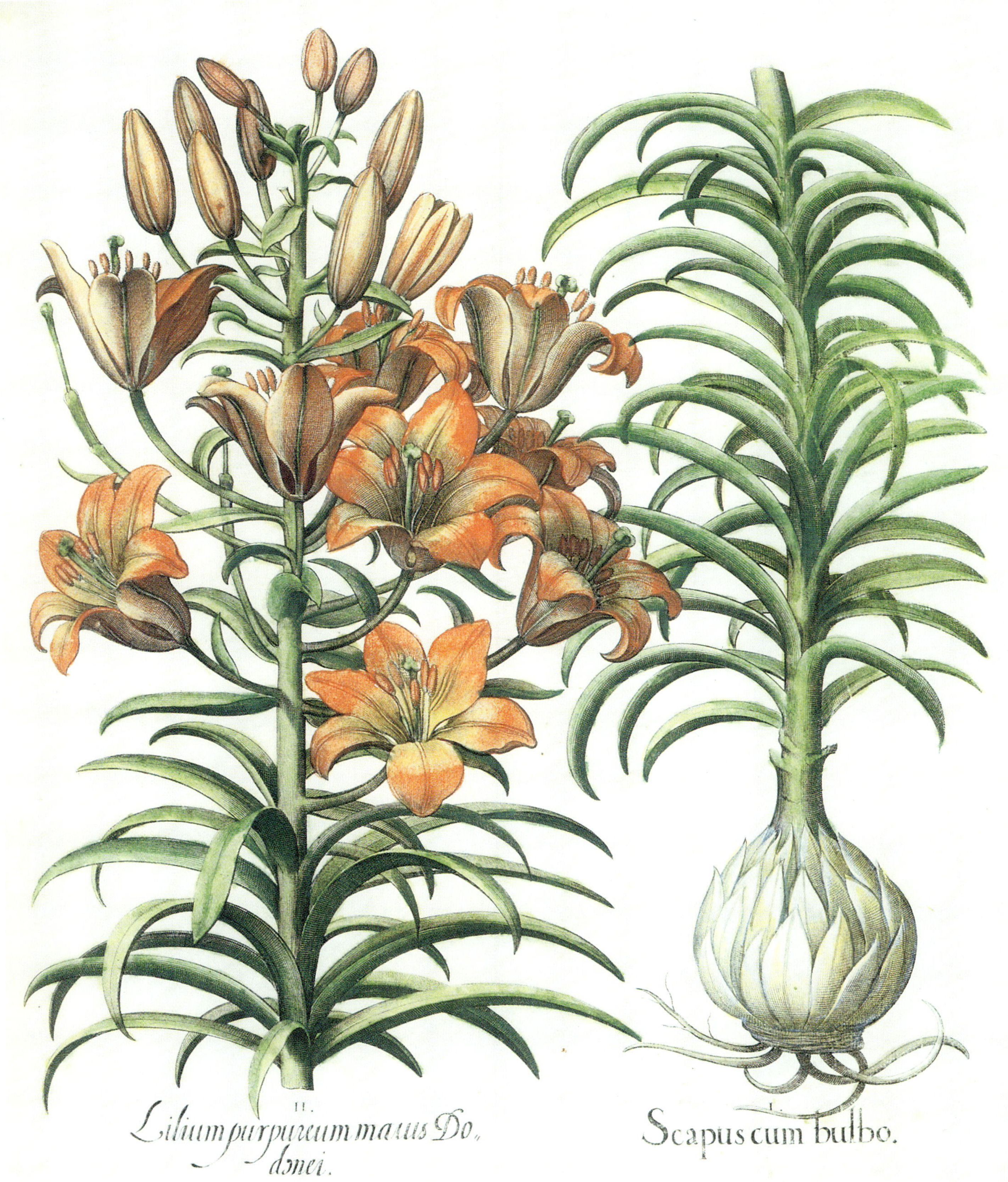

I, II. Fire lily with bulb
(Lilium bulbiferum ssp. croceum).

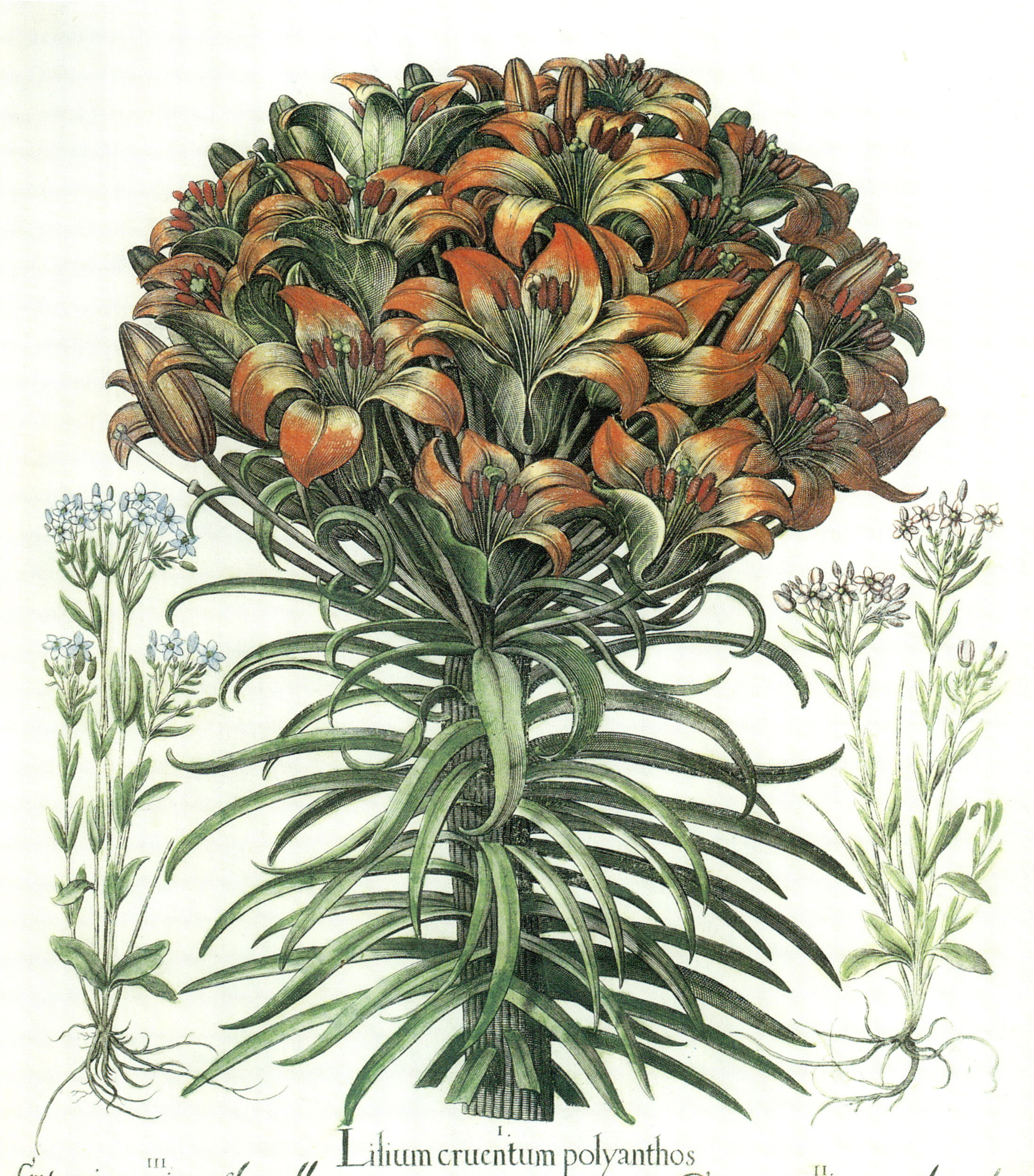

I. Proliferation of fire lily *(Lilium bulbiferum).* **II, III.** Common centaury *(Centaurium erythraea).*

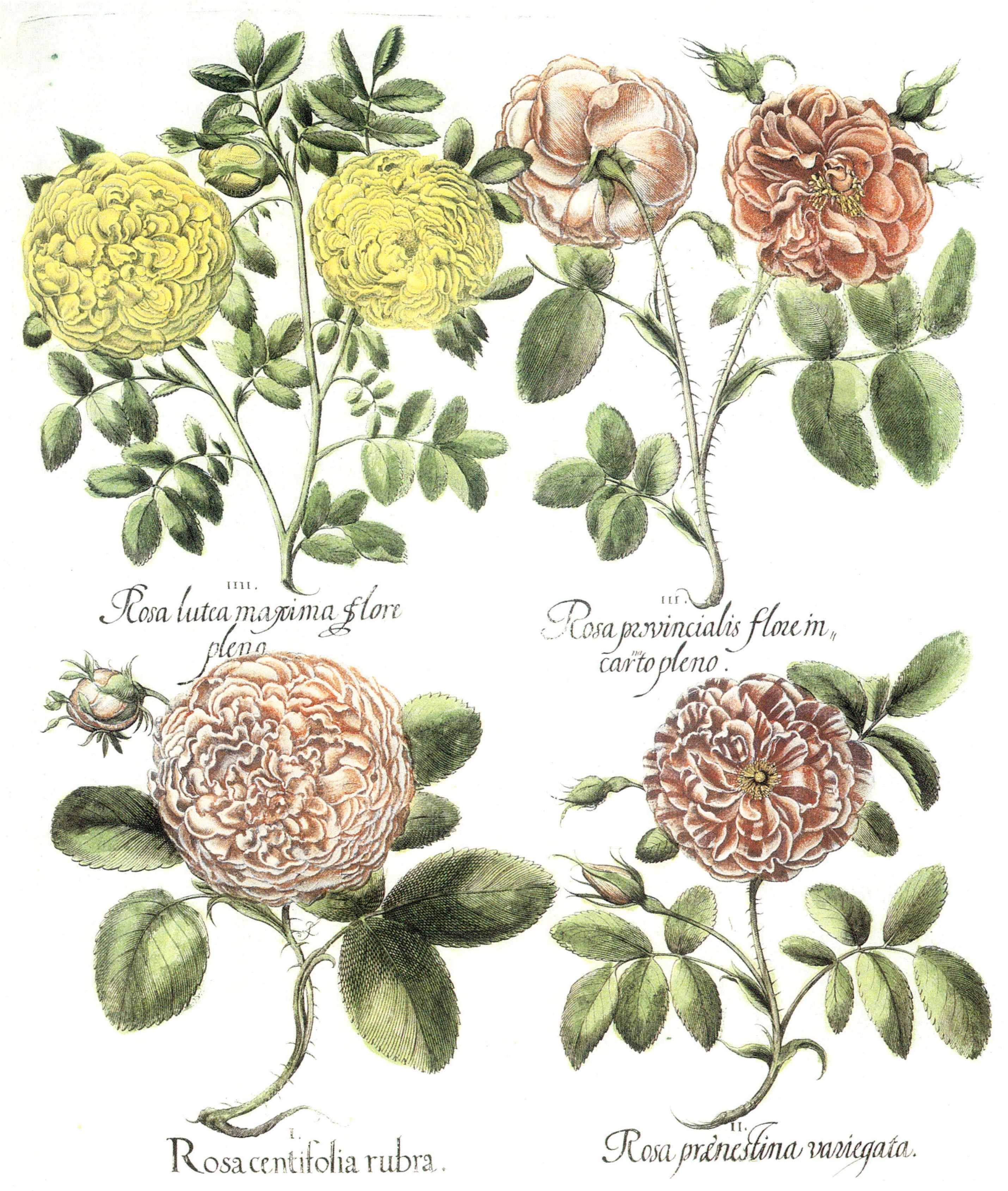

I. Provence rose *(Rosa centifolia).* **II.** Rosa mundi *(Rosa gallica officinalis-versicolor).* **III.** Apothecary's rose *(Rosa gallica officinalis).* **IV (IIII).** Sulphur rose *(Rosa hemisphaerica).*

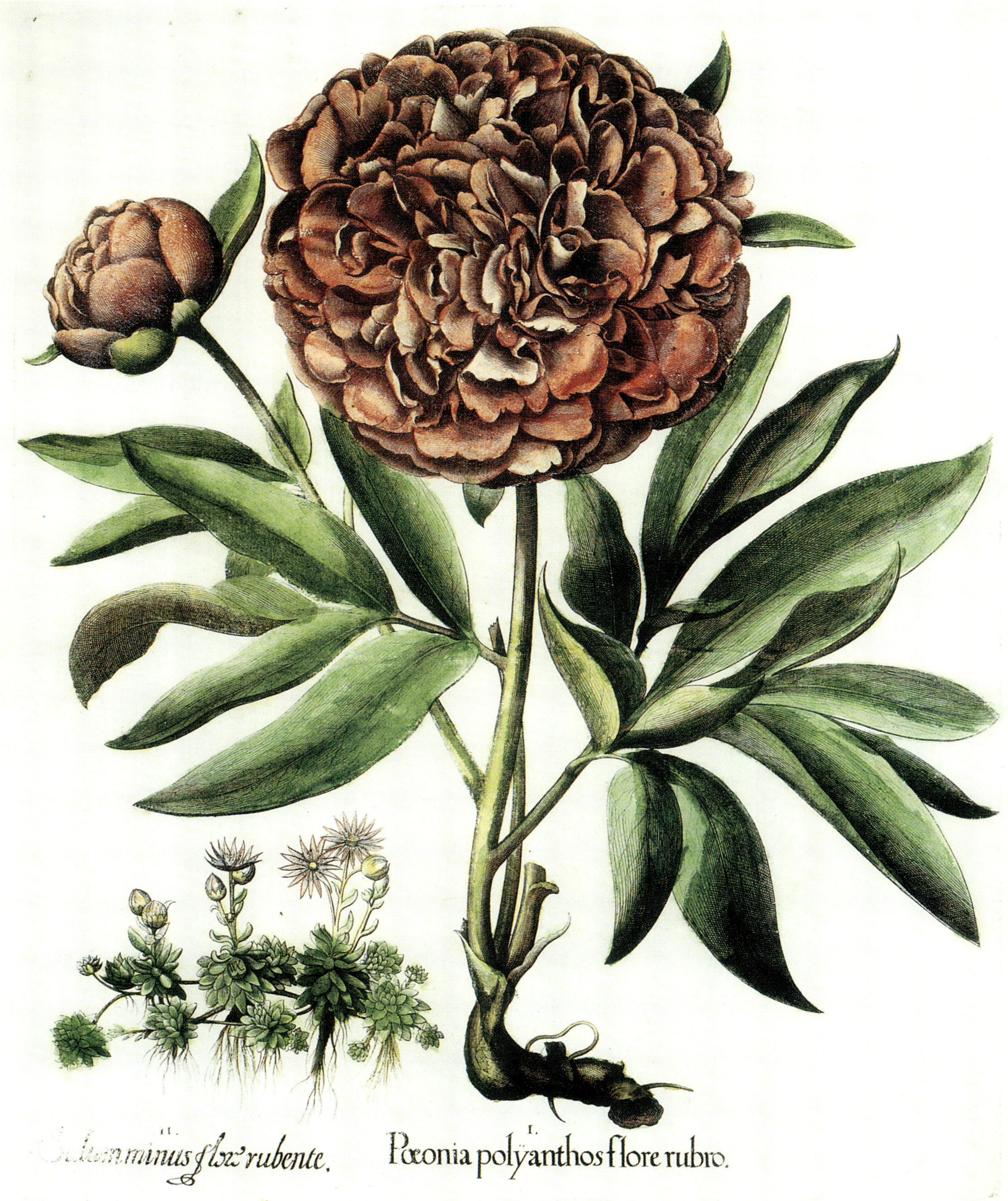

I. Double common peony *(Paeonia officinalis).* **II.** Alpine houseleek *(Sempervivum montanum).*

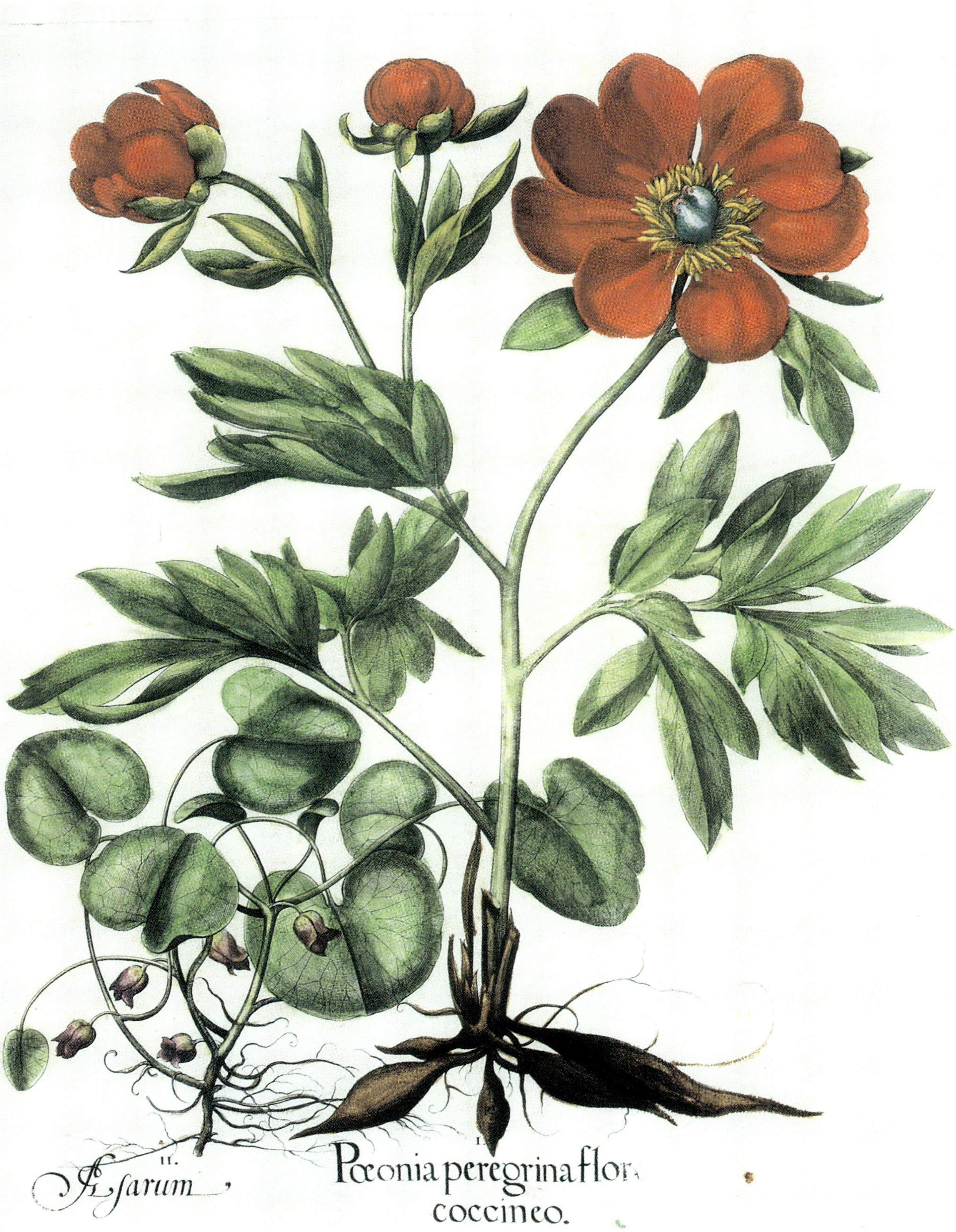

I. Single common peony *(Paeonia officinalis).*
II. Asarabacca *(Asarum europaeum).*

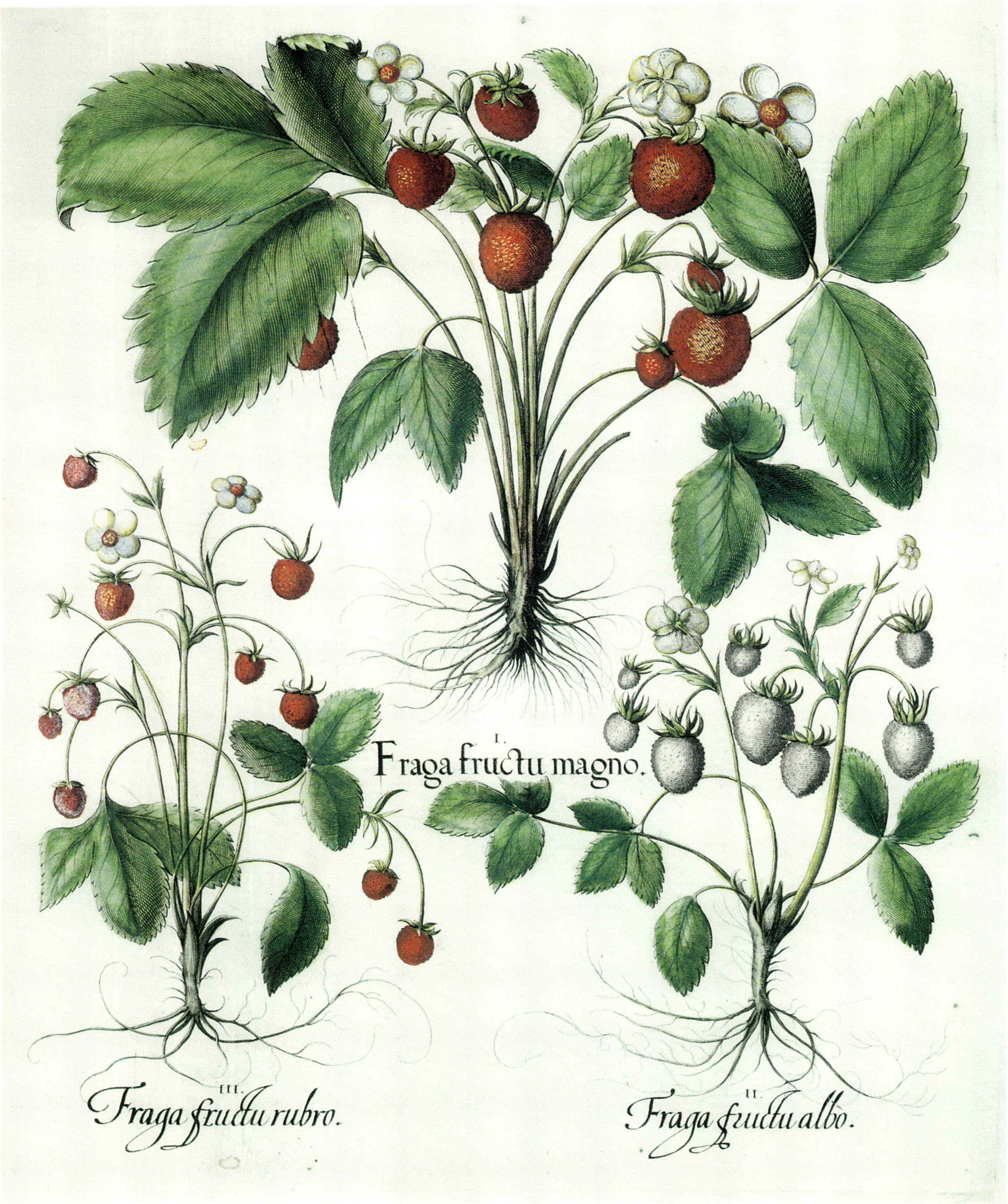

I. Large-fruited strawberry *(Fragaria spec.)*. **II.** White-fruited strawberry *(Fragaria spec.)*. **III.** Small-fruited strawberry *(Fragaria spec.)*.

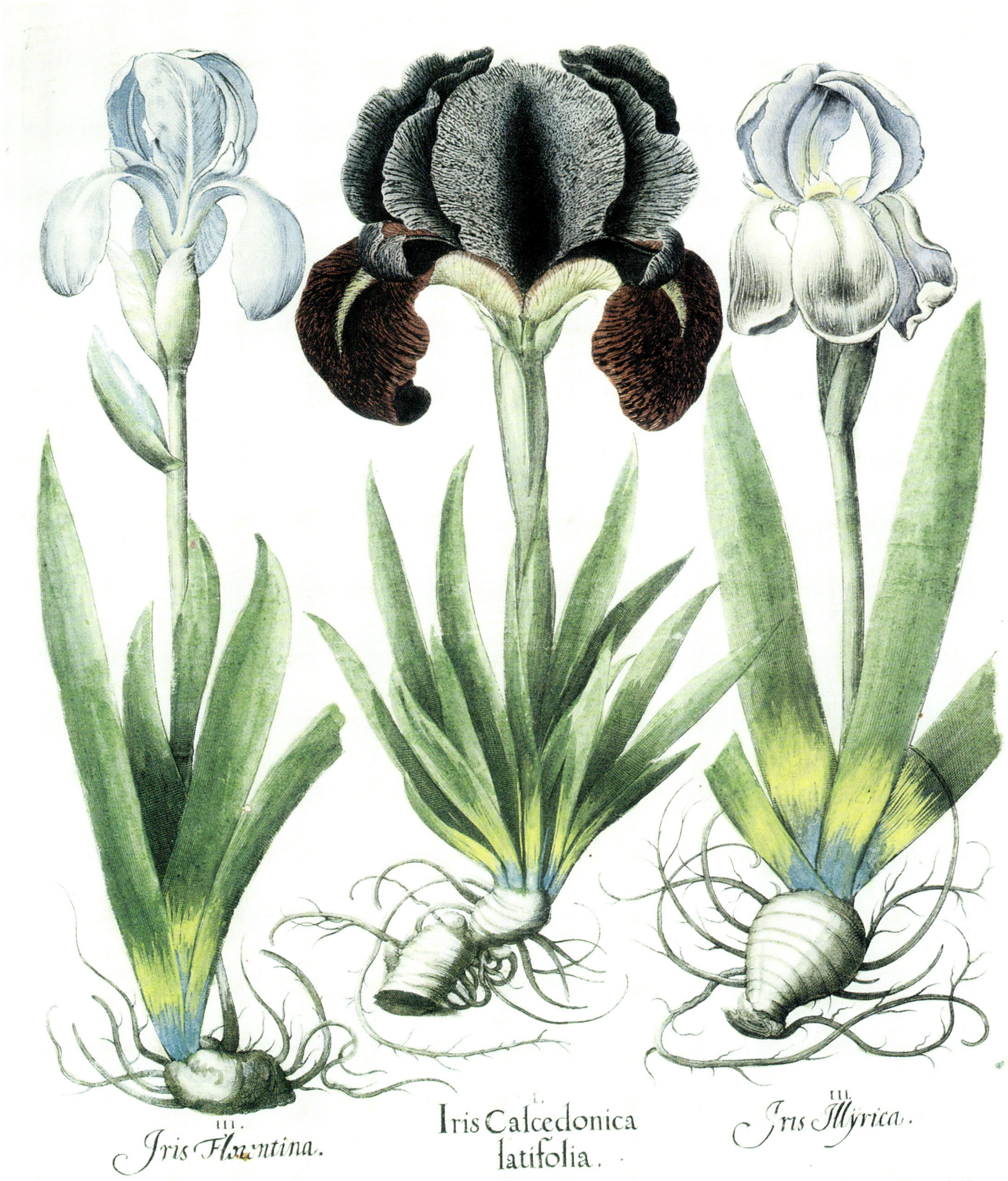

I. Iris *(Iris spec.)*. **II.** Orris root *(Iris florentina)*.
III. Dalmatian iris *(Iris pallida)*.

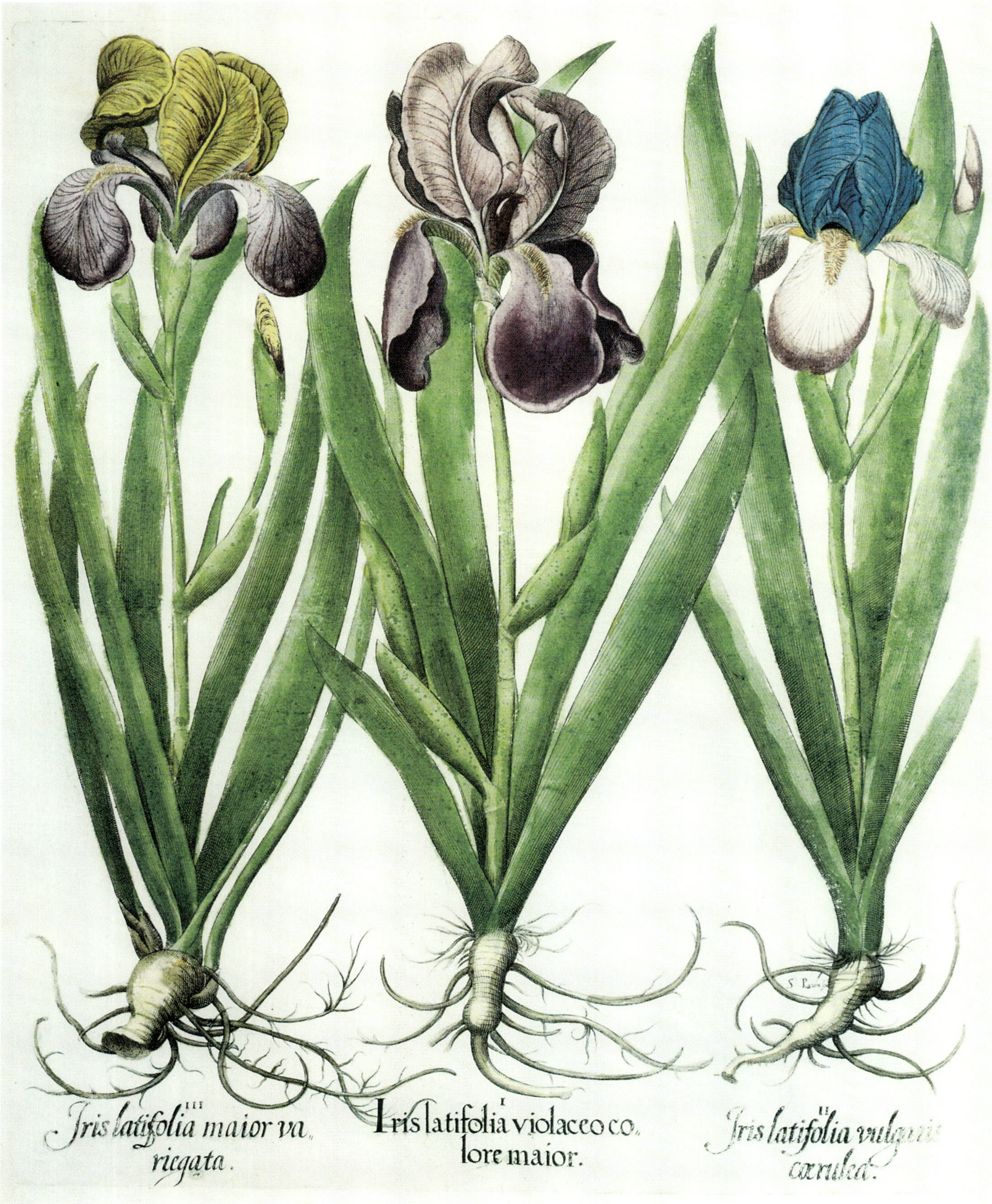

I, II, III. German iris
(Iris germanica).

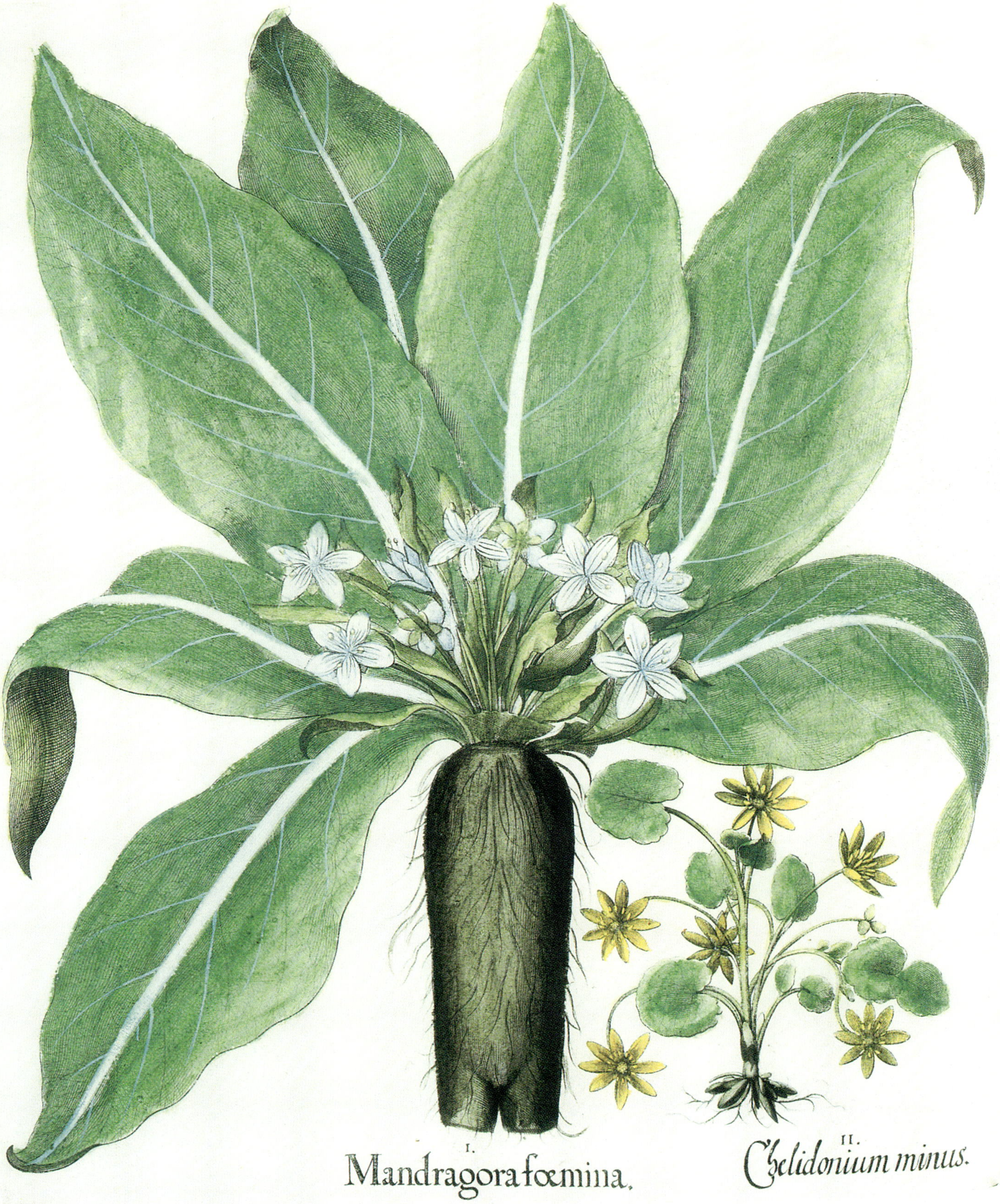

I. Autumn mandrake *(Mandragora autumnalis)*.
II. Lesser celandine *(Ranunculus ficaria)*.

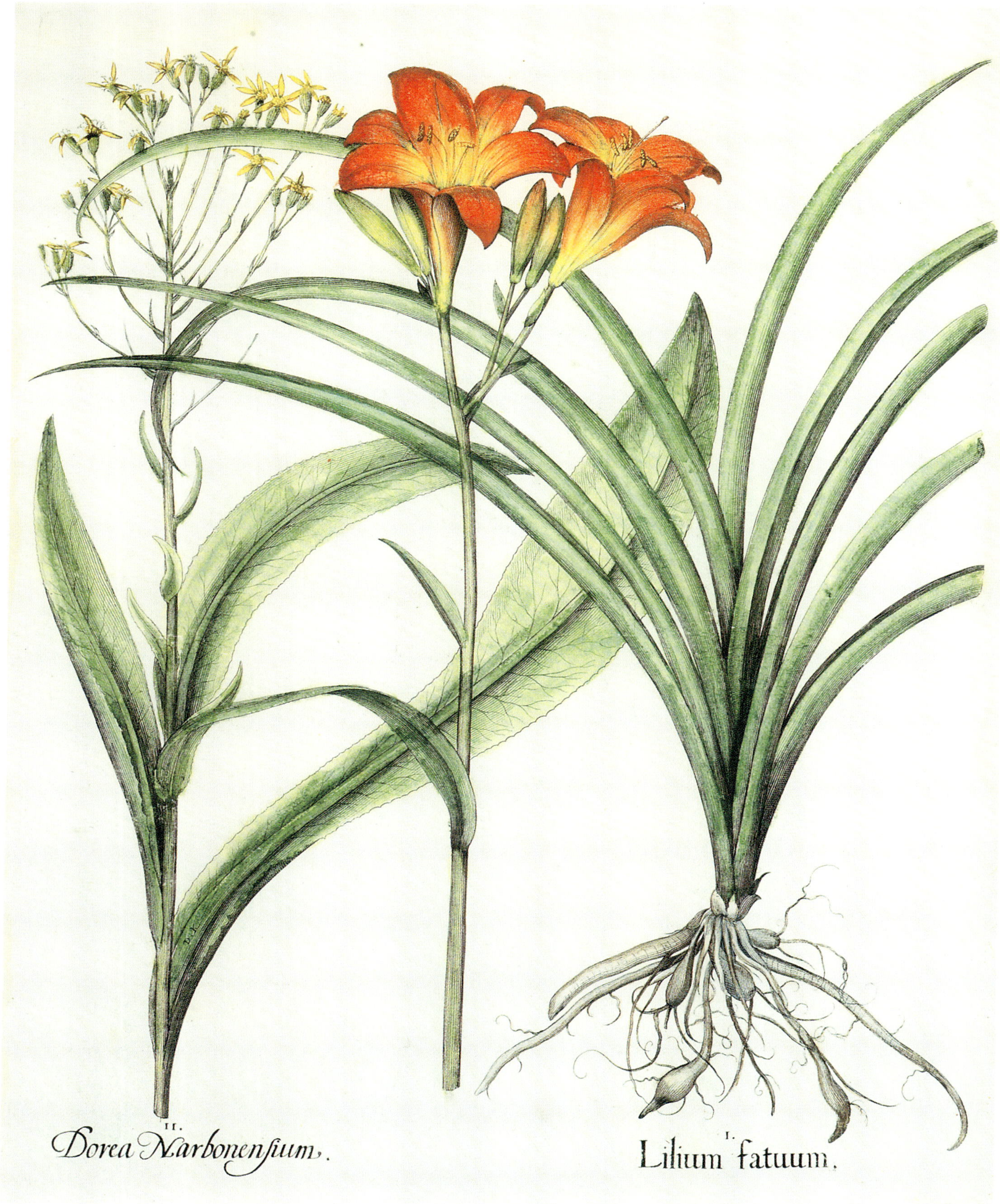

I. Red-yellow day lily *(Hemerocallis fulva)*.
II. Groundsel *(Senecio doria)*.

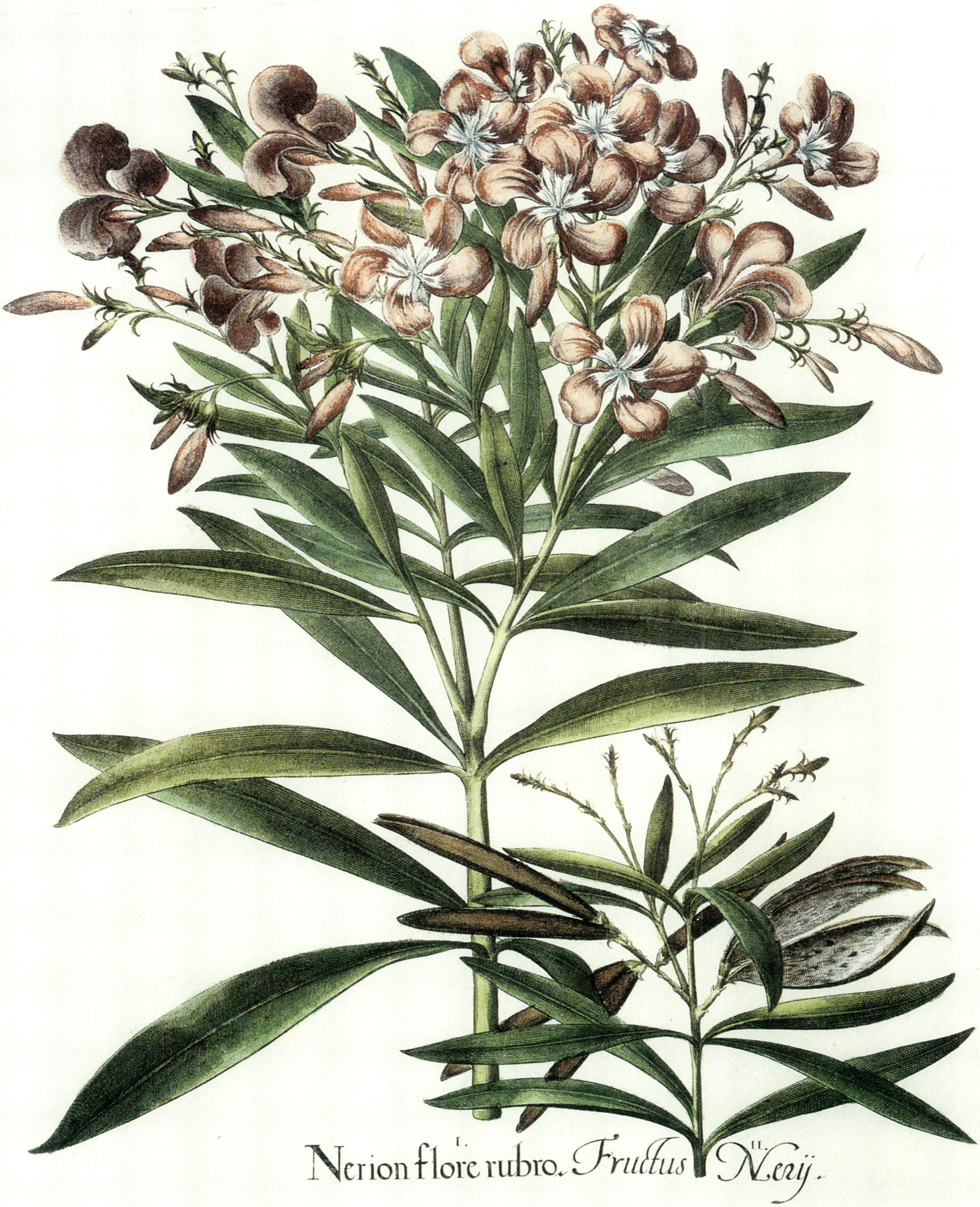

I. Rose bay; Oleander *(Nerium oleander).* **II.** Rose bay; Oleander (in fruit) *(Nerium oleander).*

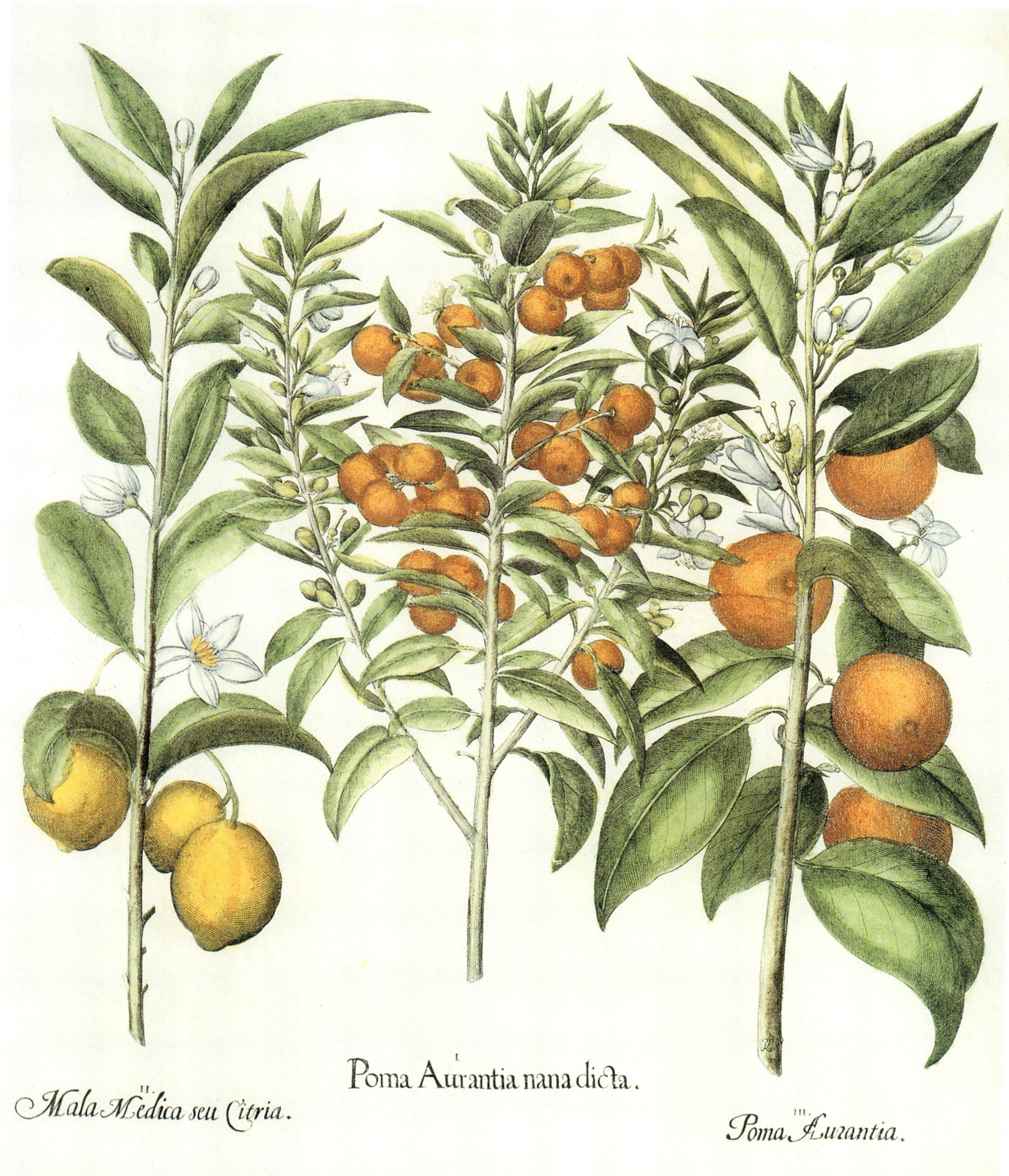

I. Seville orange; Bitter orange *(Citrus aurantium)*. **II.** Citron *(Citrus medica)*.
III. Common or sweet orange *(Citrus sinensis)*.

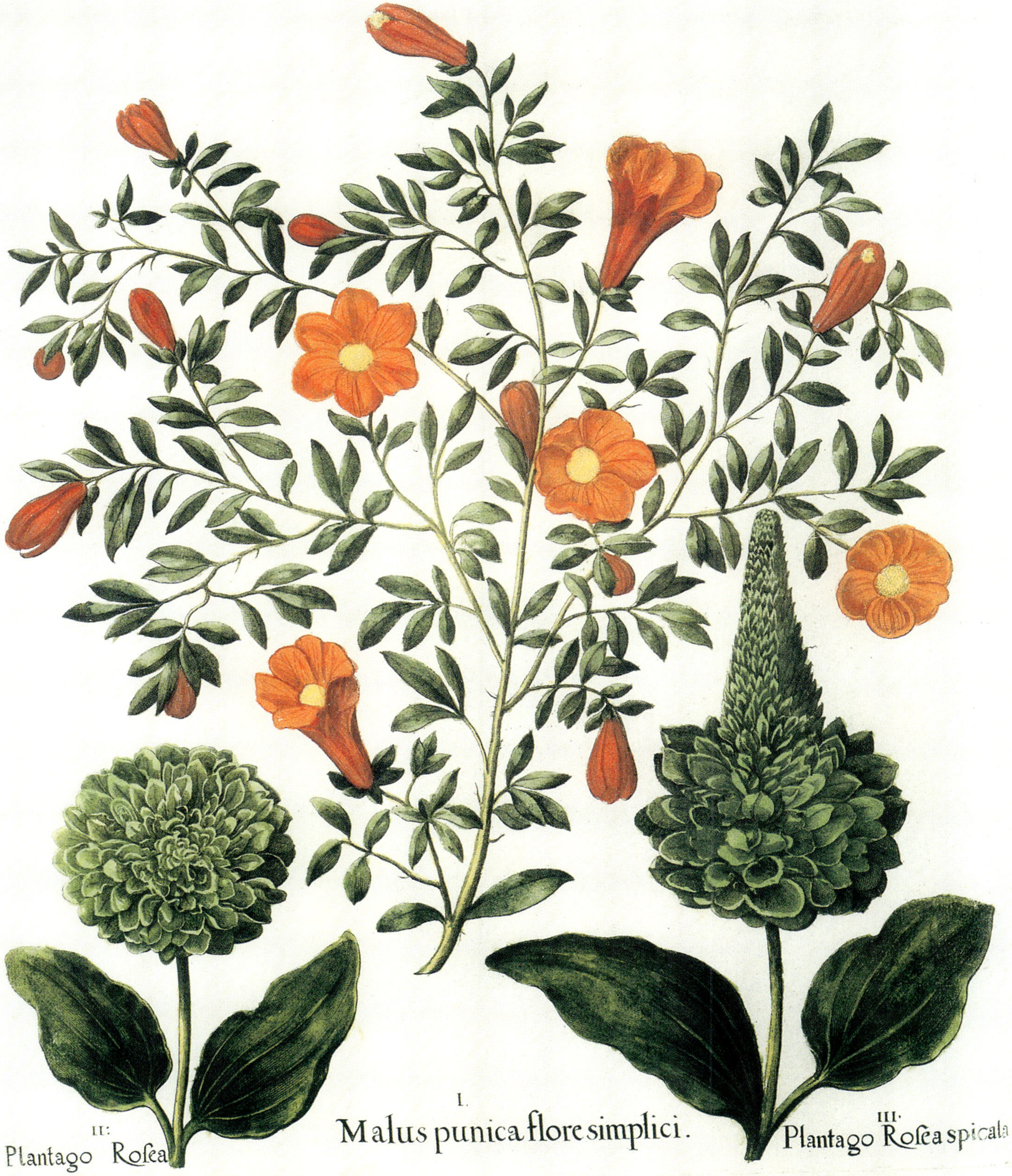

I. Pomegranate *(Punica granatum).* **II, III.** Common plantain *(Plantago major).*

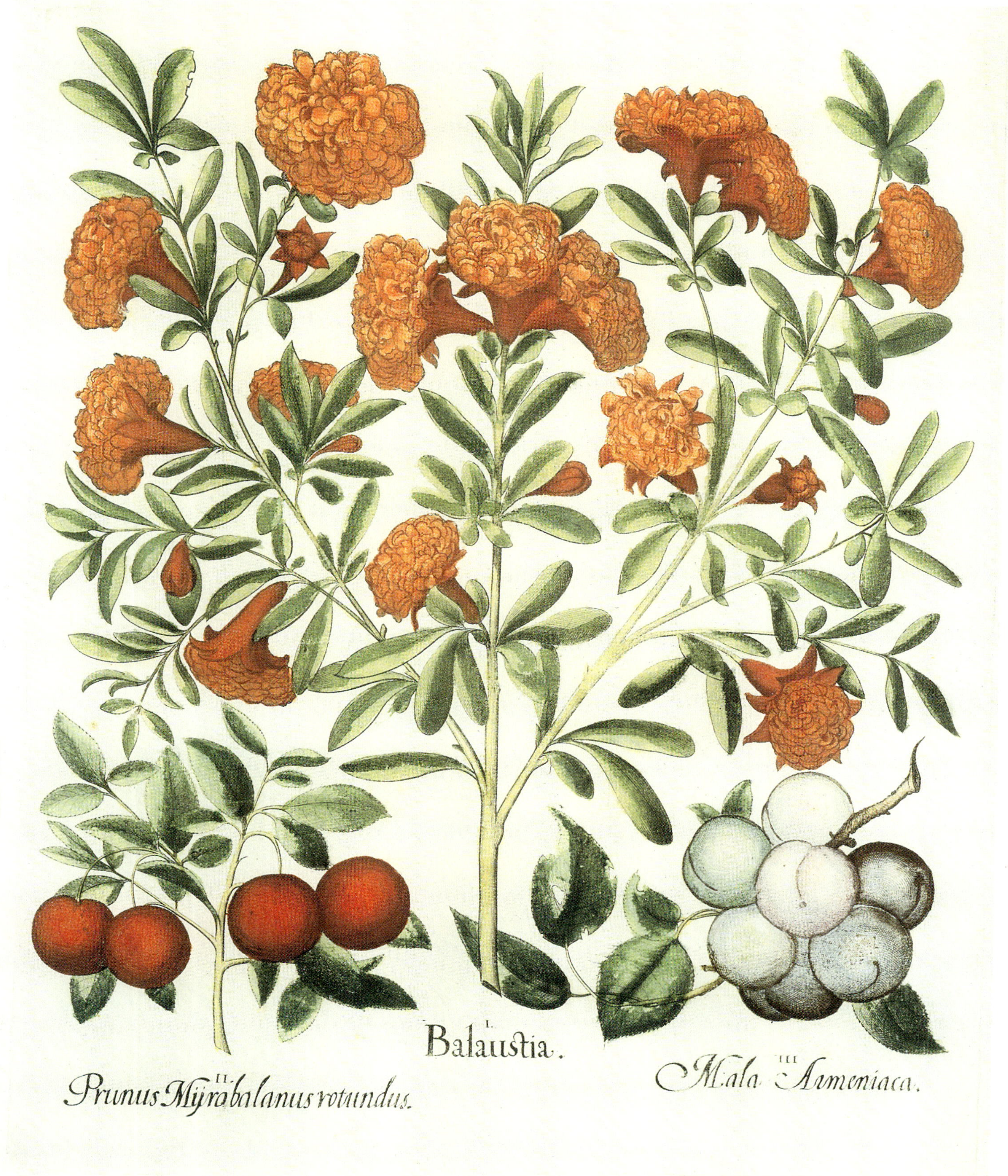

I. Pomegranate with fruits *(Punica granatum)*. **II.** Cherry plum *(Prunus cerasifera)*. **III.** Apricot *(Prunus armeniaca)*.

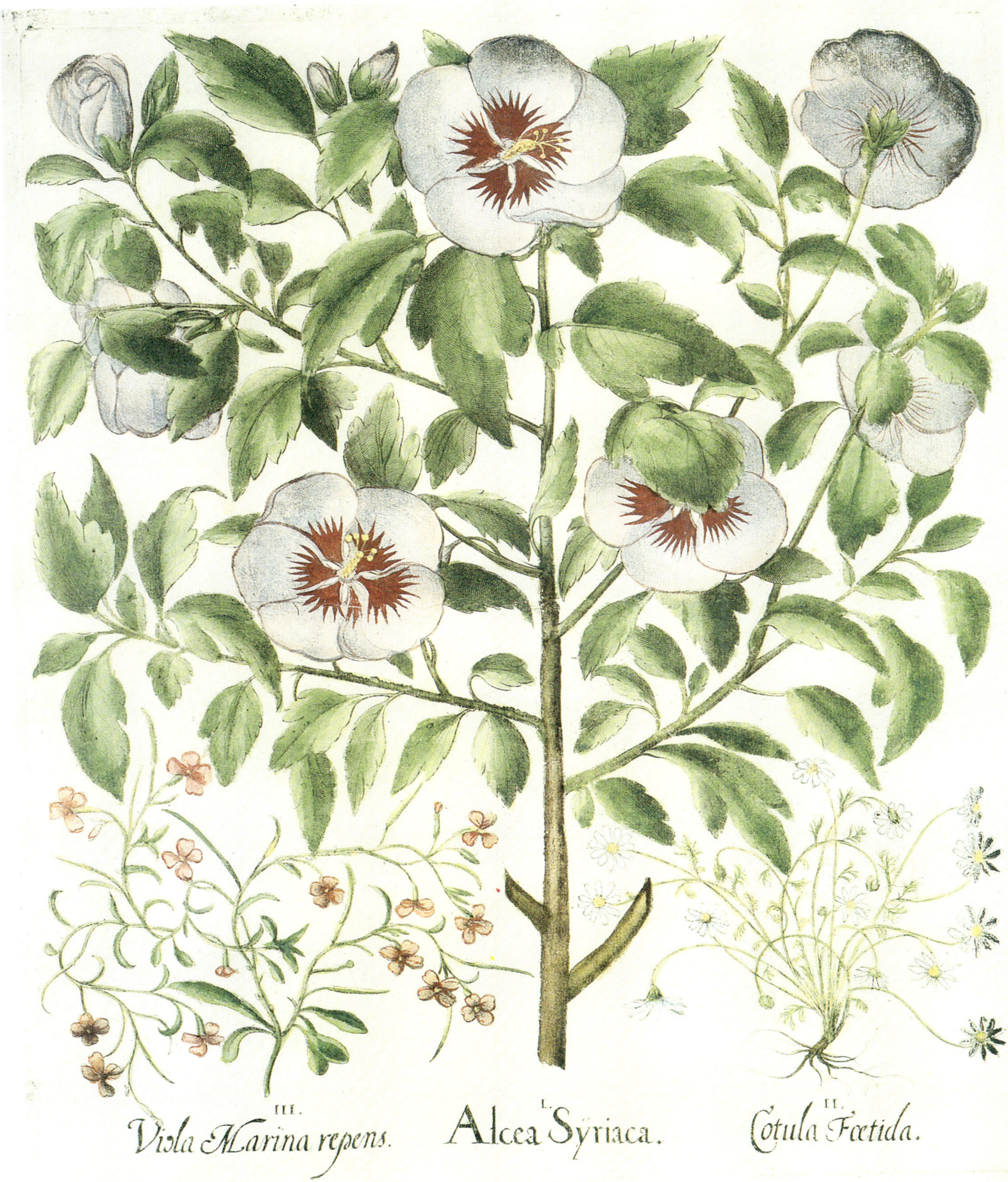

I. Mallow *(Hibiscus syriacus).* **II.** Mayweed *(Anthemis cotula).*
III. Virginia stock *(Malcolmia maritima).*

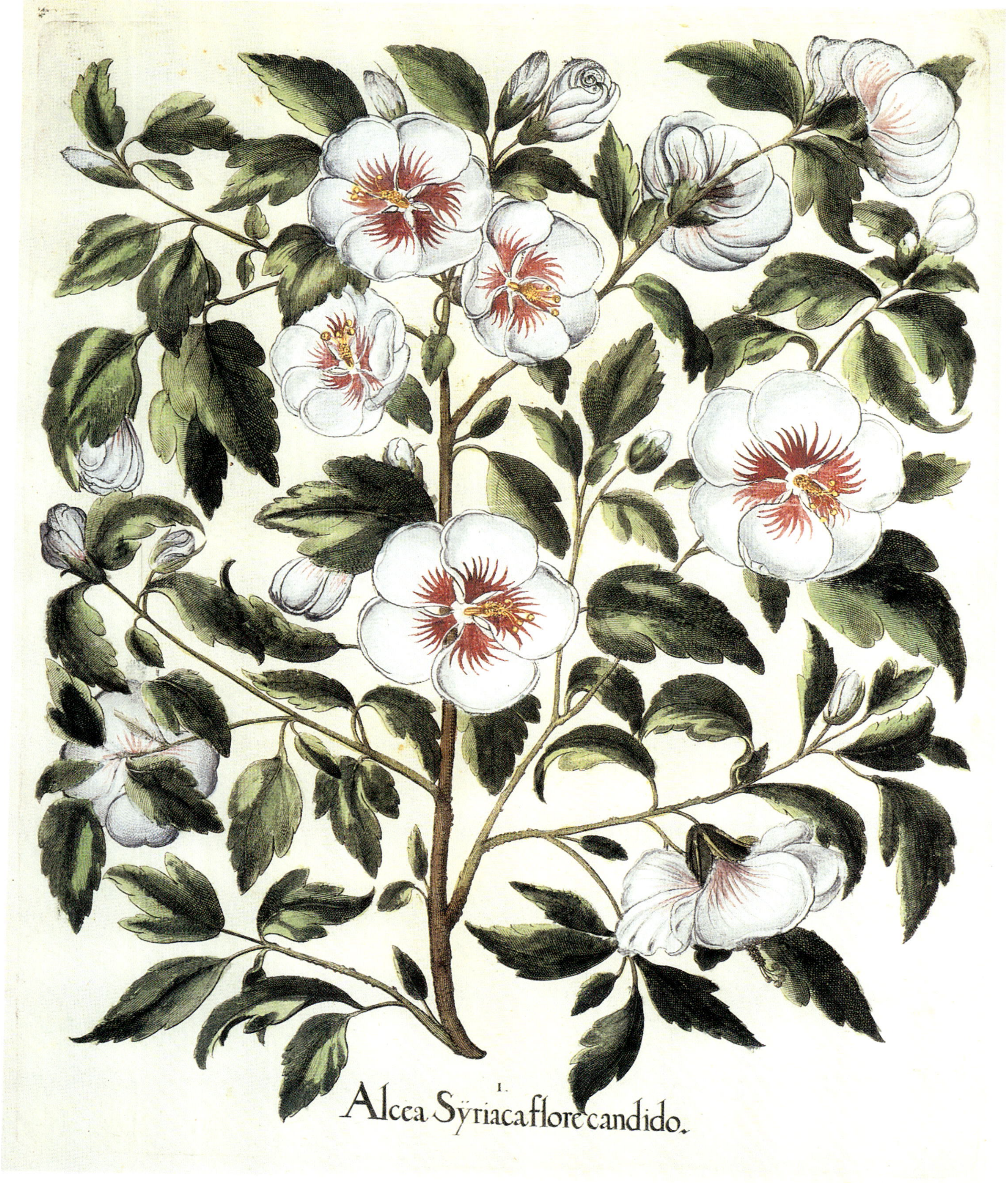

I. White-flowering mallow; Rose mallow
(Hibiscus syriacus).

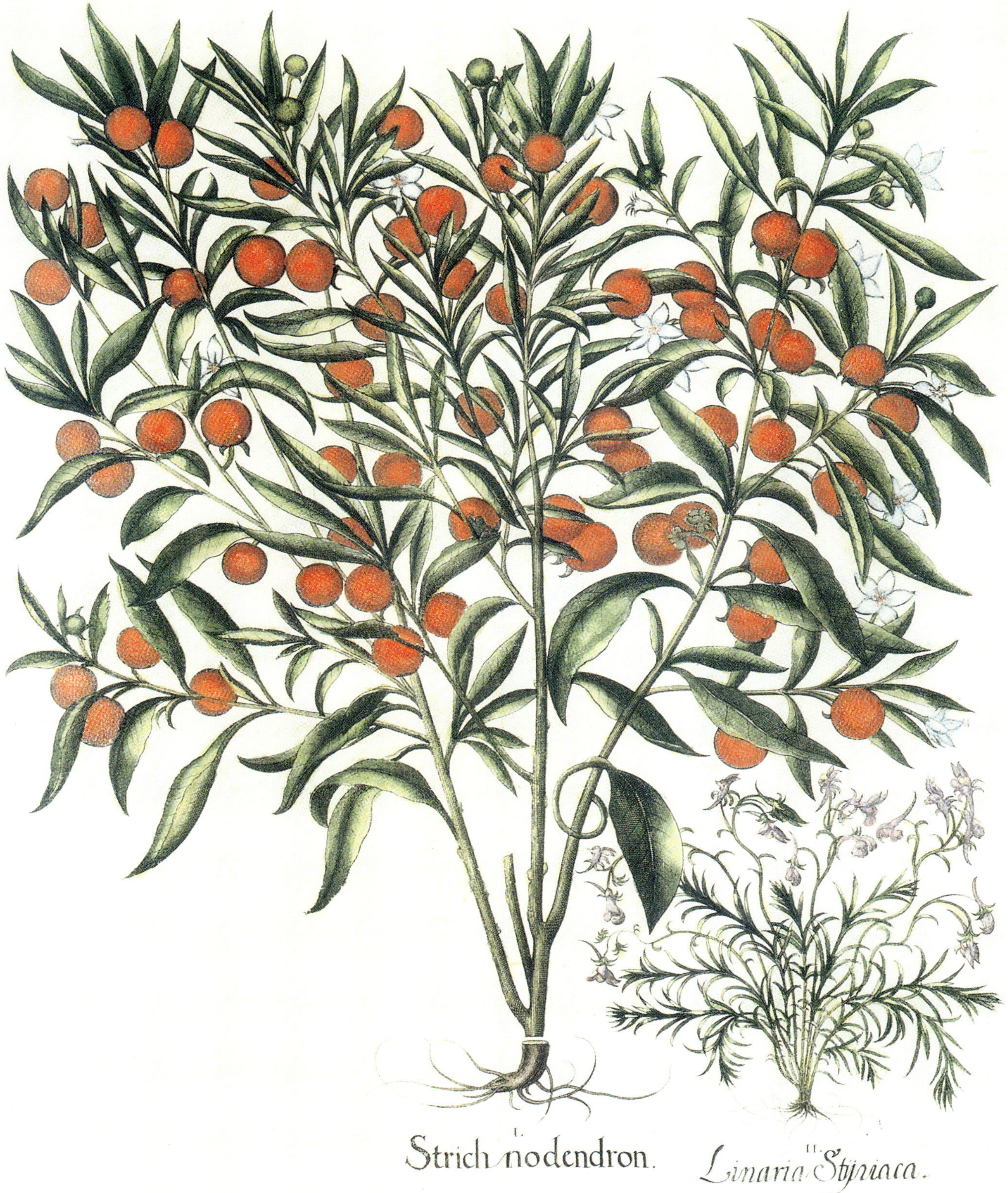

I. Winter cherry; Jerusalem cherry *(Solanum pseudocapsicum).*
II. Alpine toadflax *(Linaria alpina).*

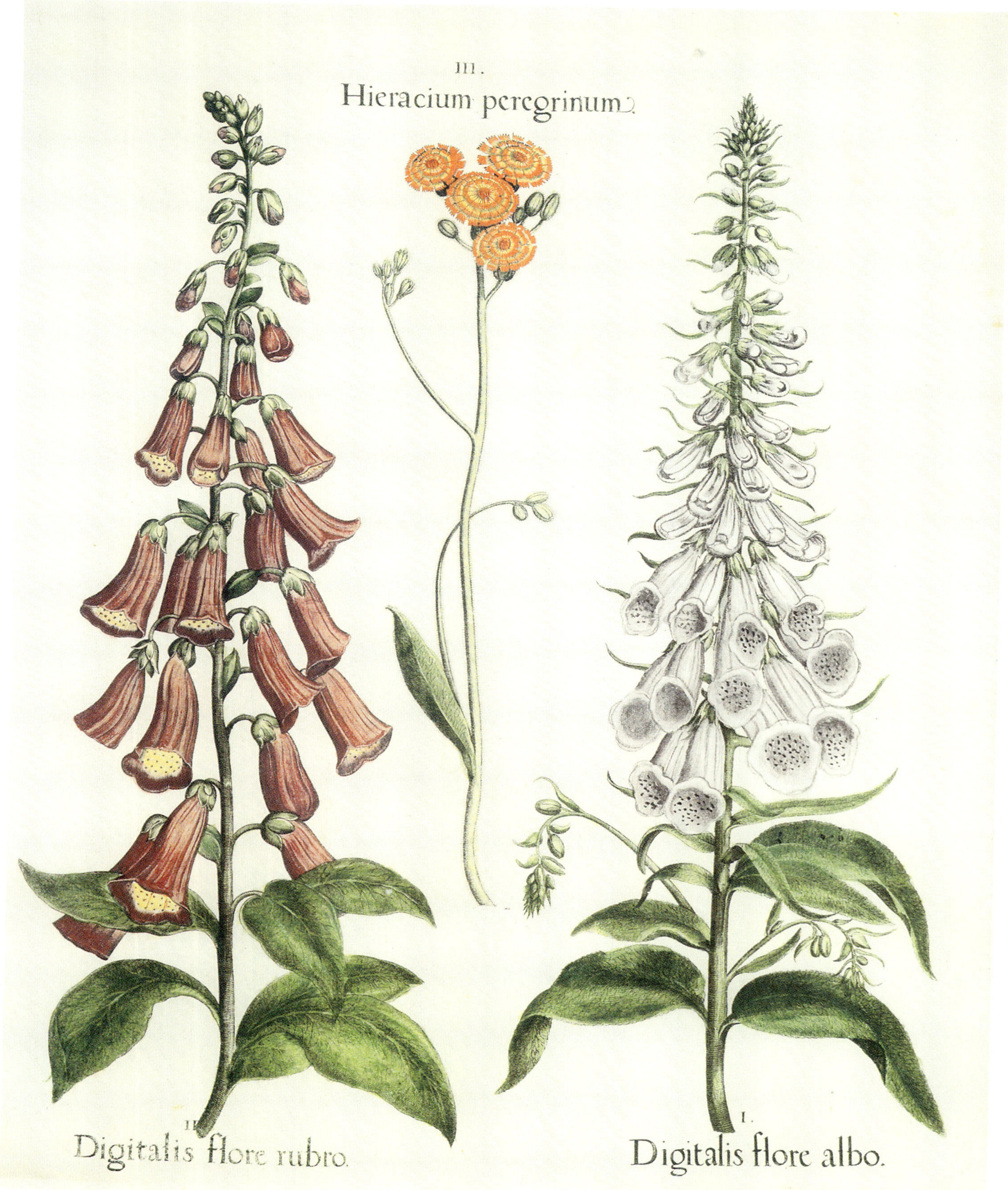

I. White foxglove *(Digitalis purpurea)*. **II.** Red foxglove *(Digitalis purpurea)*.
III. Orange hawkweed *(Hieracium aurantiacum)*.

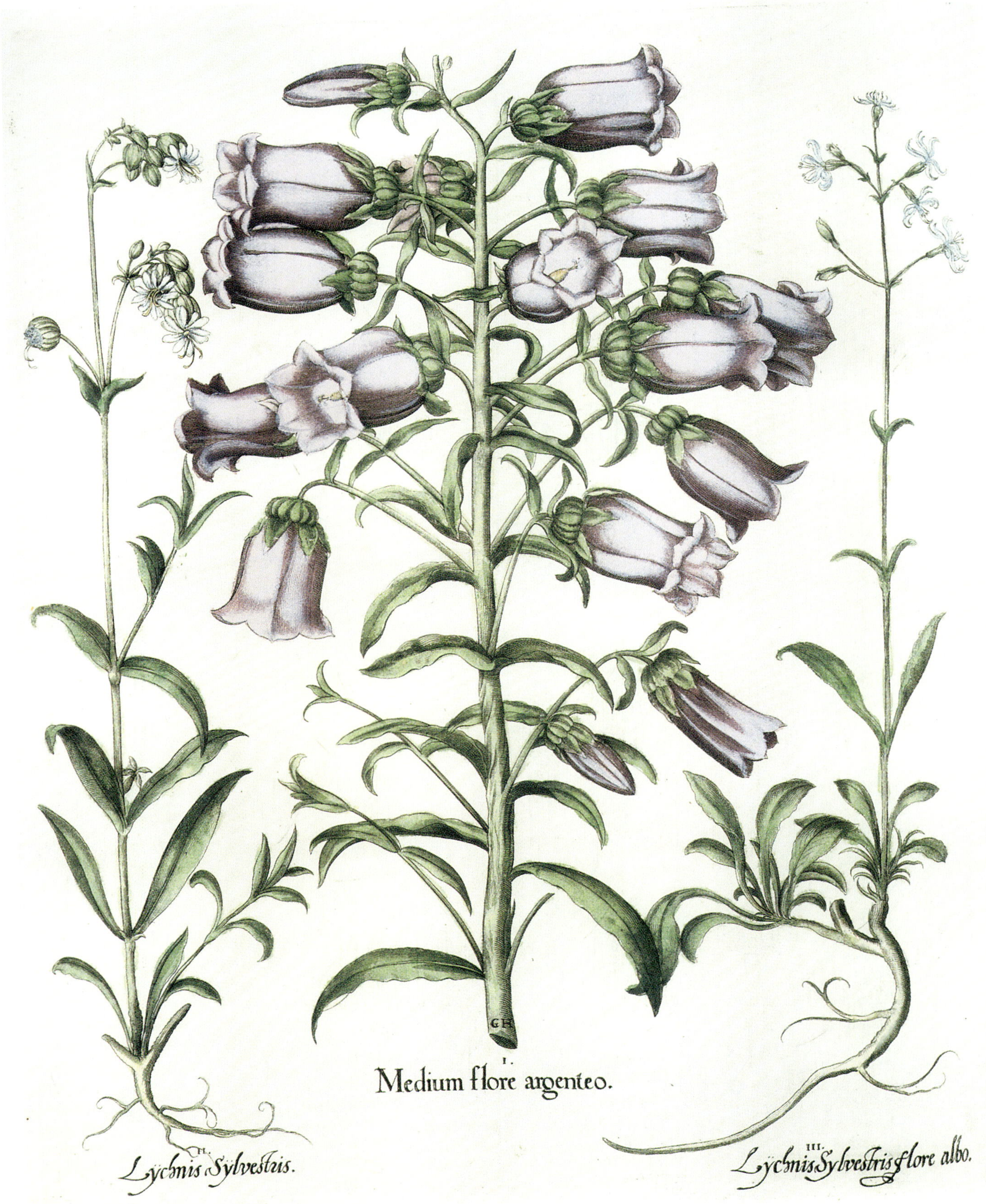

I. Canterbury bells *(Campanula medium)*. **II.** Bladder campion *(Silene vulgaris)*.
III. Campion *(Silene nutans)*.

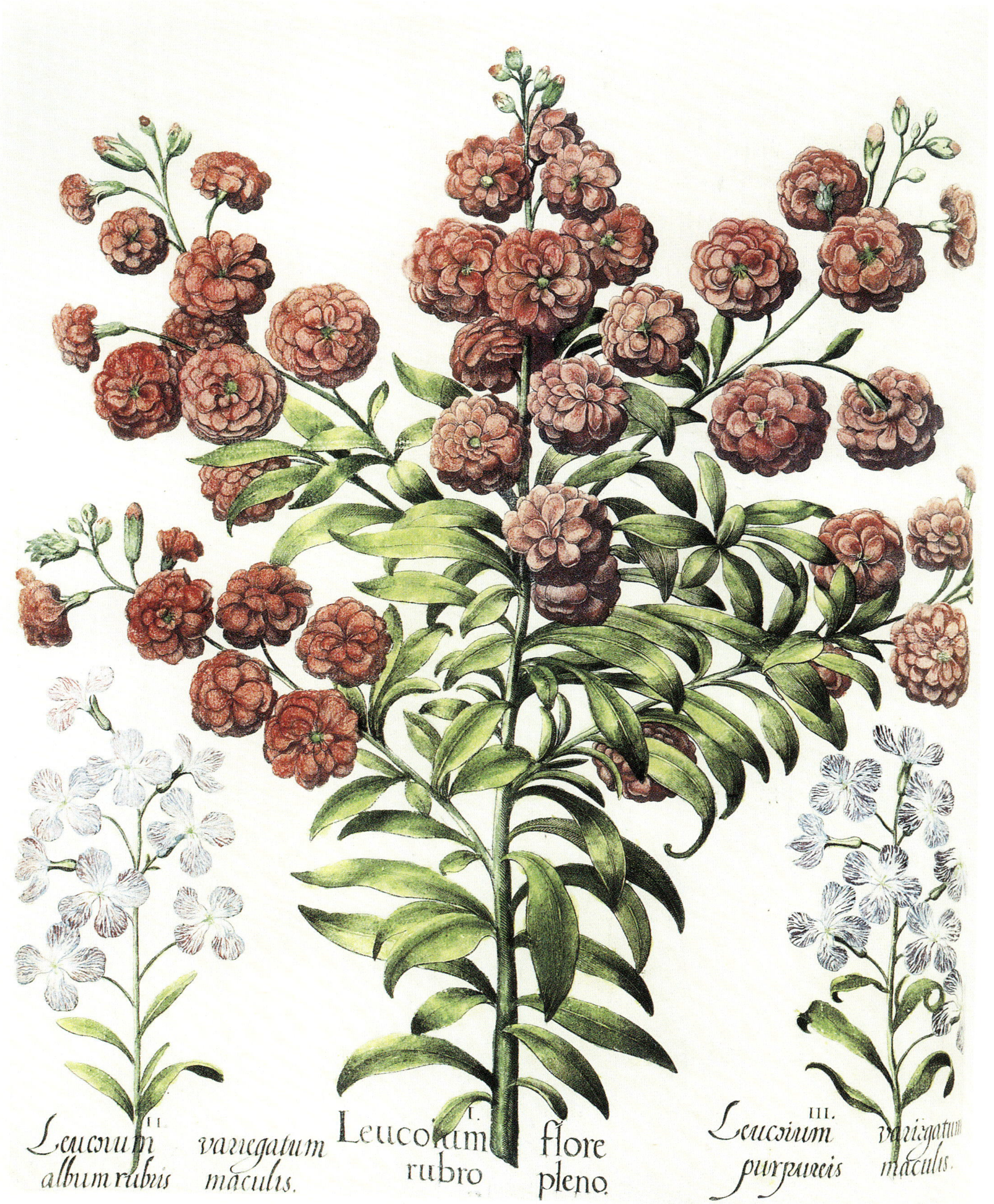

I. Double pink gillyflower *(Matthiola incana).* **II, III.** Gillyflower *(Matthiola incana).*

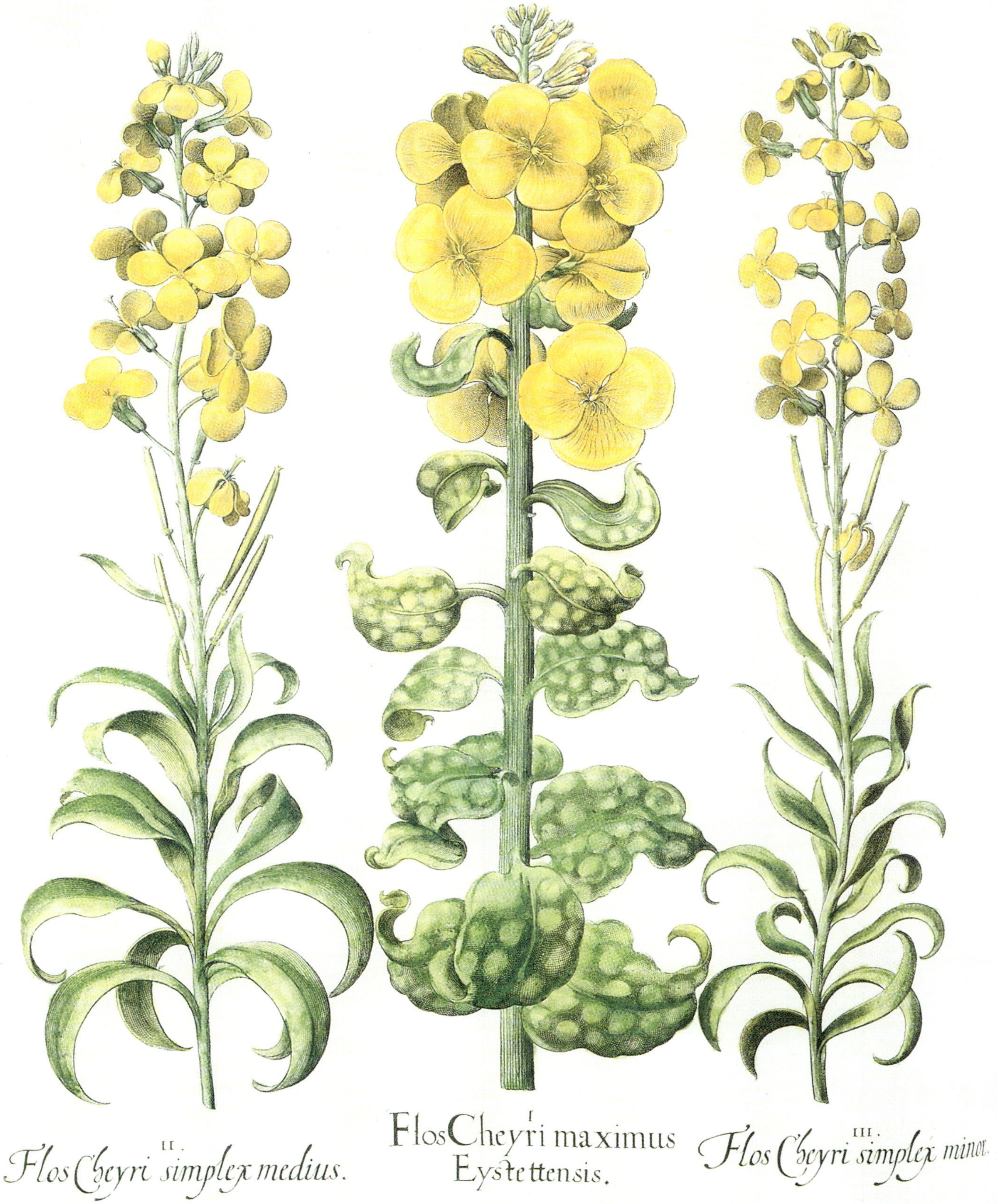

I. "Wallflower of Eichstätt" *(Cheiranthus cheiri).* **II, III.** Single wallflower *(Cheiranthus cheiri).*

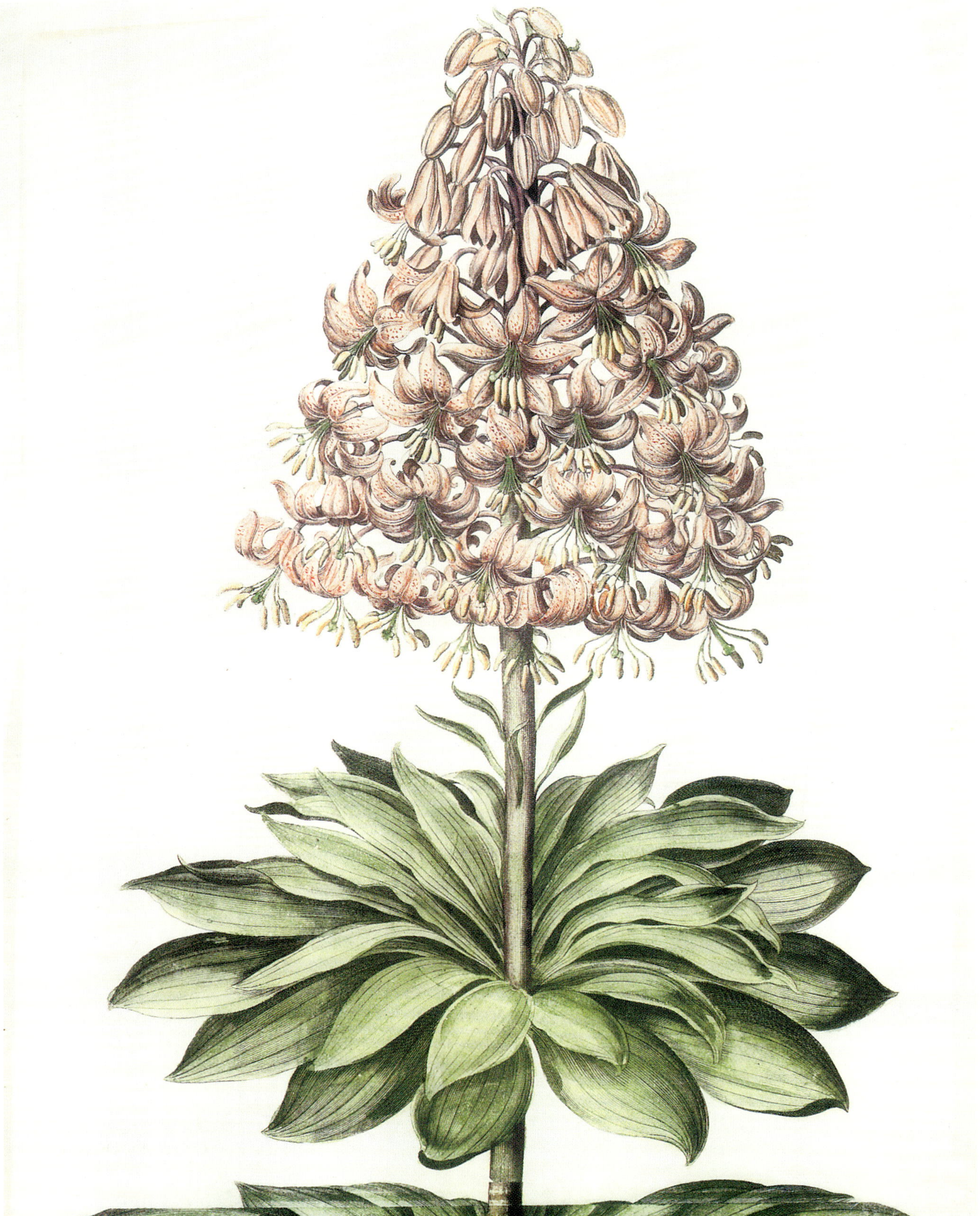

Turk's-cap lily
(Lilium martagon)

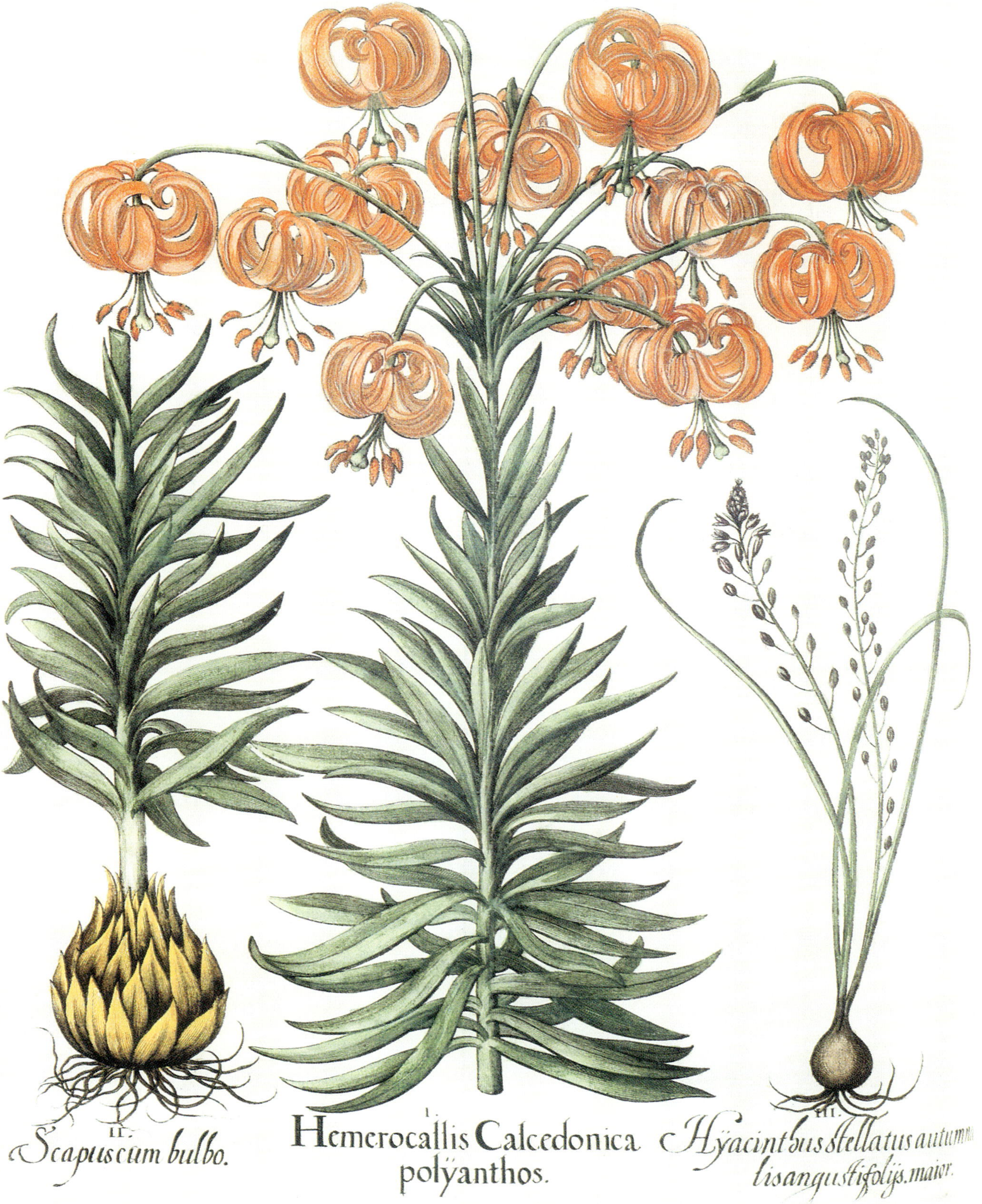

I, II. Scarlet turk's-cap lily *(Lilium chalcedonicum).*
III. Autumn squill *(Scilla autumnalis).*

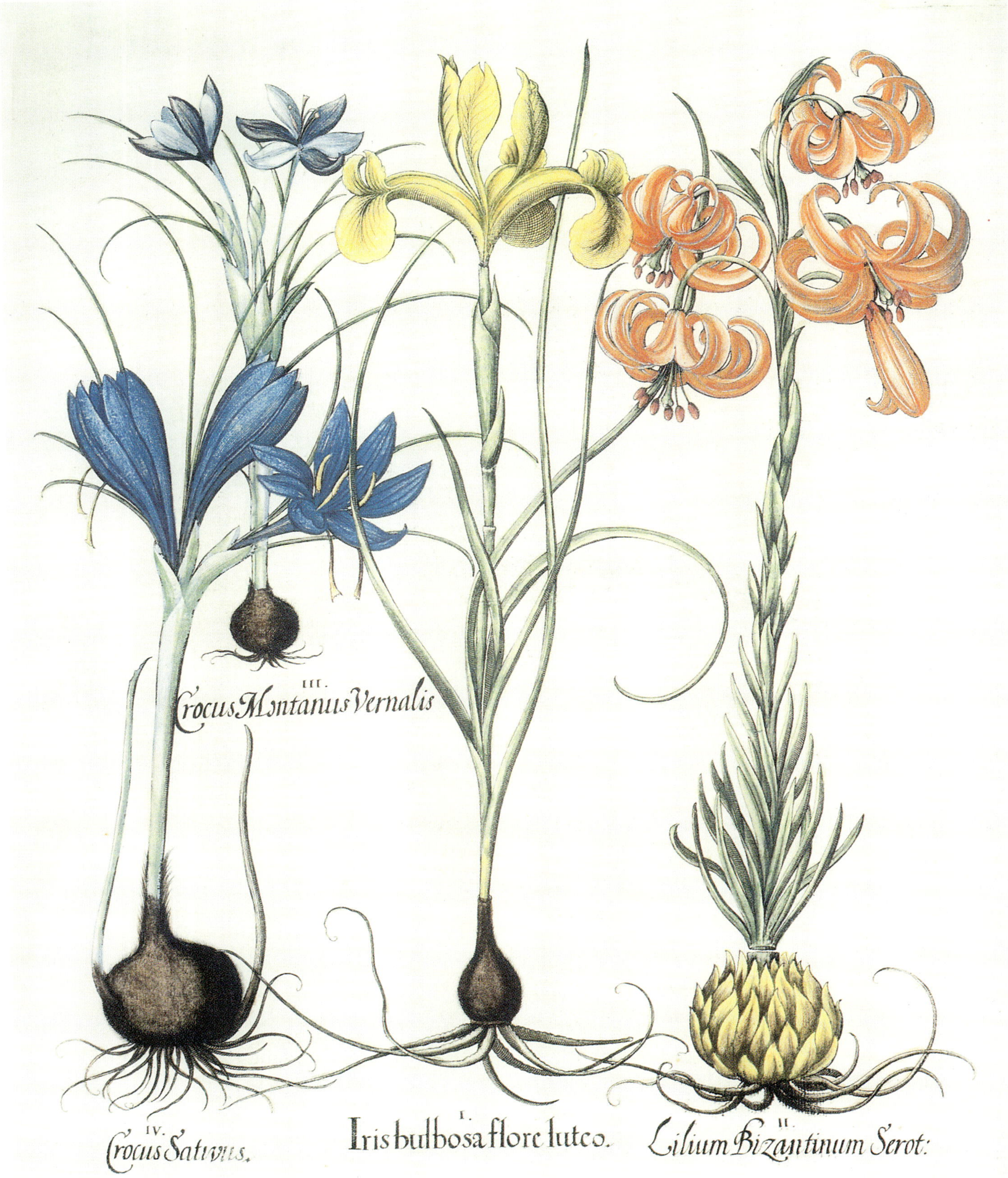

I. Spanish iris *(Iris xiphium).* **II.** Scarlet turk's-cap lily *(Lilium chalcedonicum).*
III. Dutch crocus *(Crocus spec.).* **IV.** Saffron crocus *(Crocus sativus).*

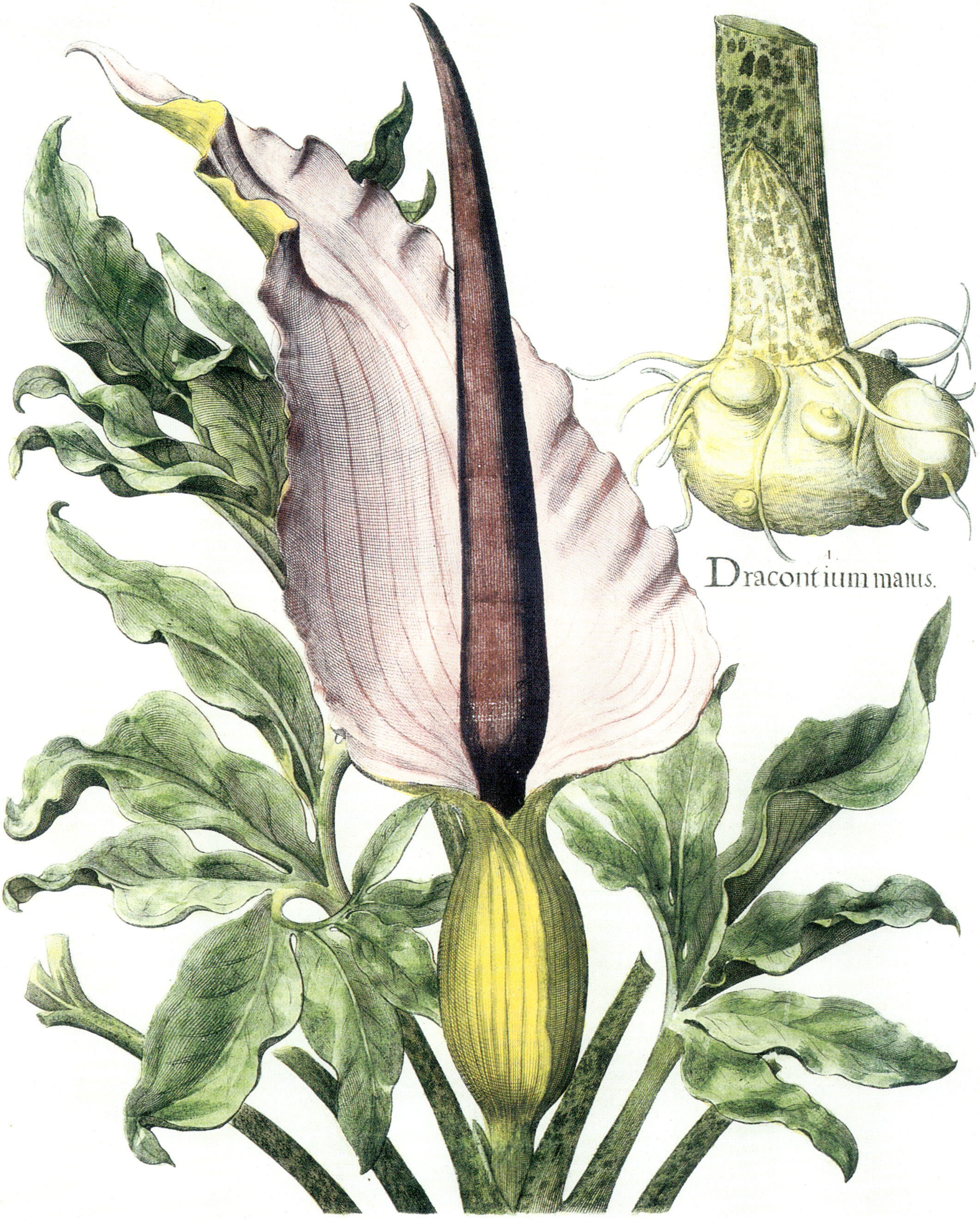

I. Dragon arum with rhizome
(Dracunculus vulgaris).

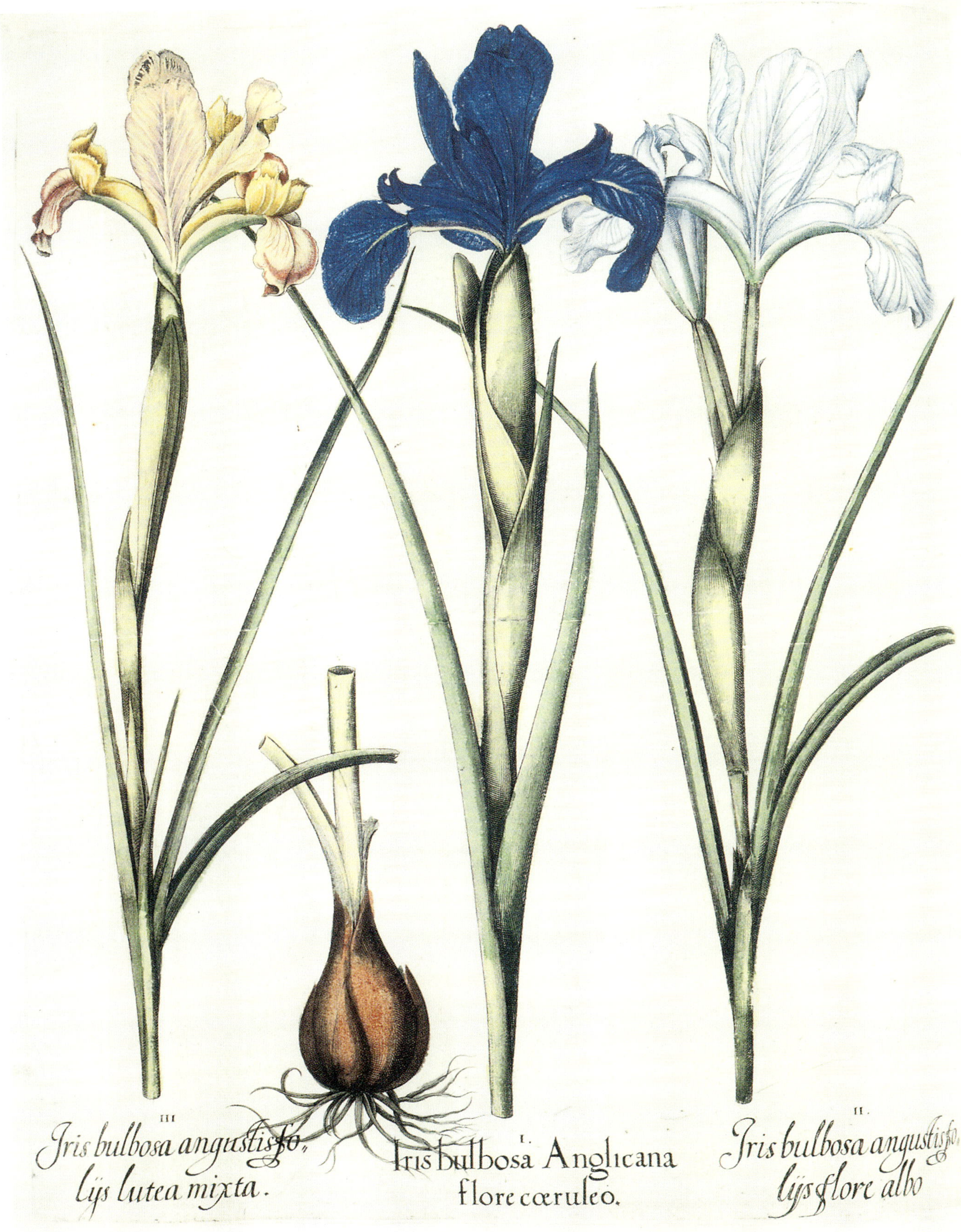

I. English iris *(Iris xiphioides).* **II.** White iris with narrow leaves *(Iris xiphium).*
III. Yellowish iris with narrow leaves *(Iris xiphium).*

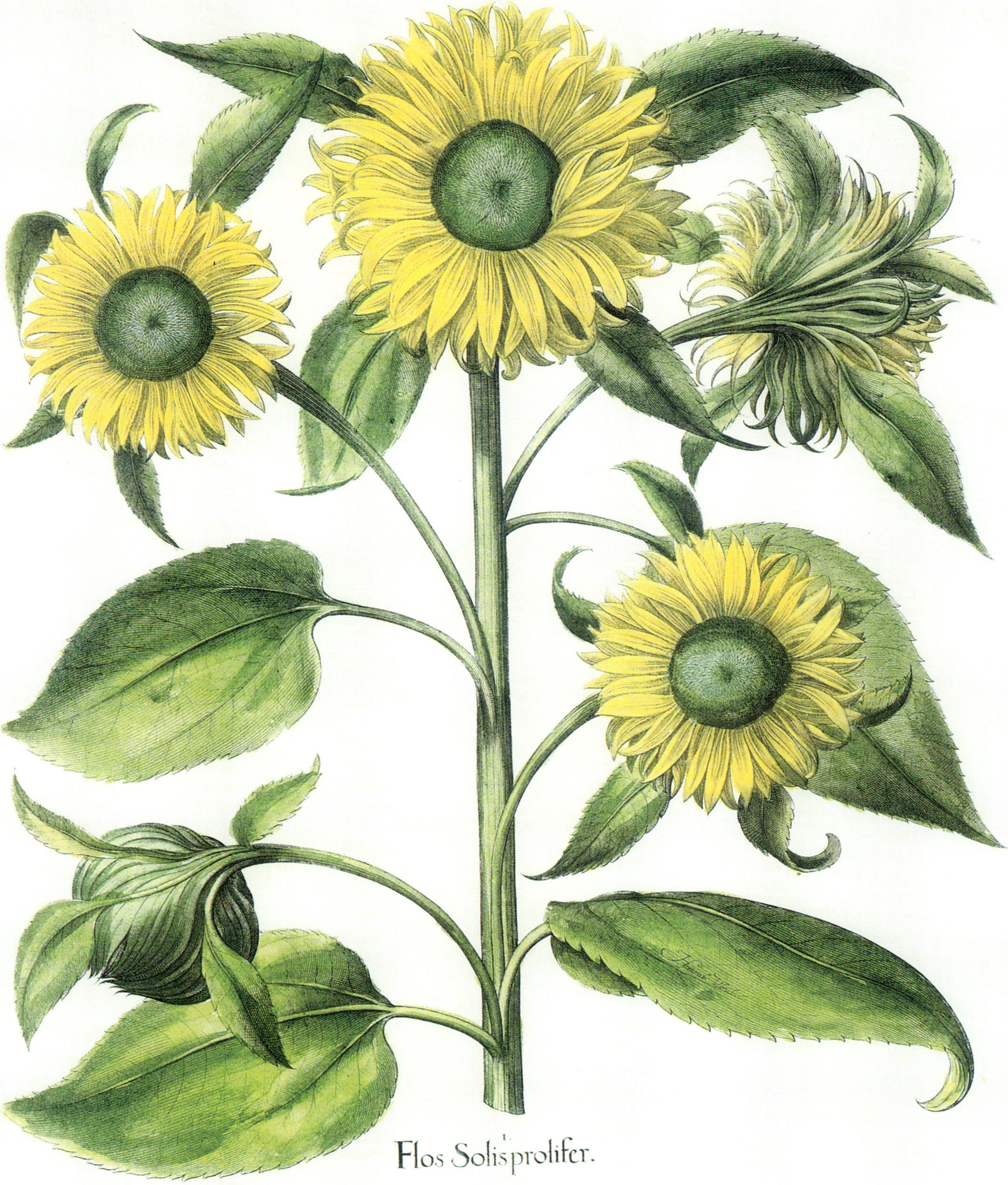

I. Sunflower
(Helianthus x *multiflorus).*

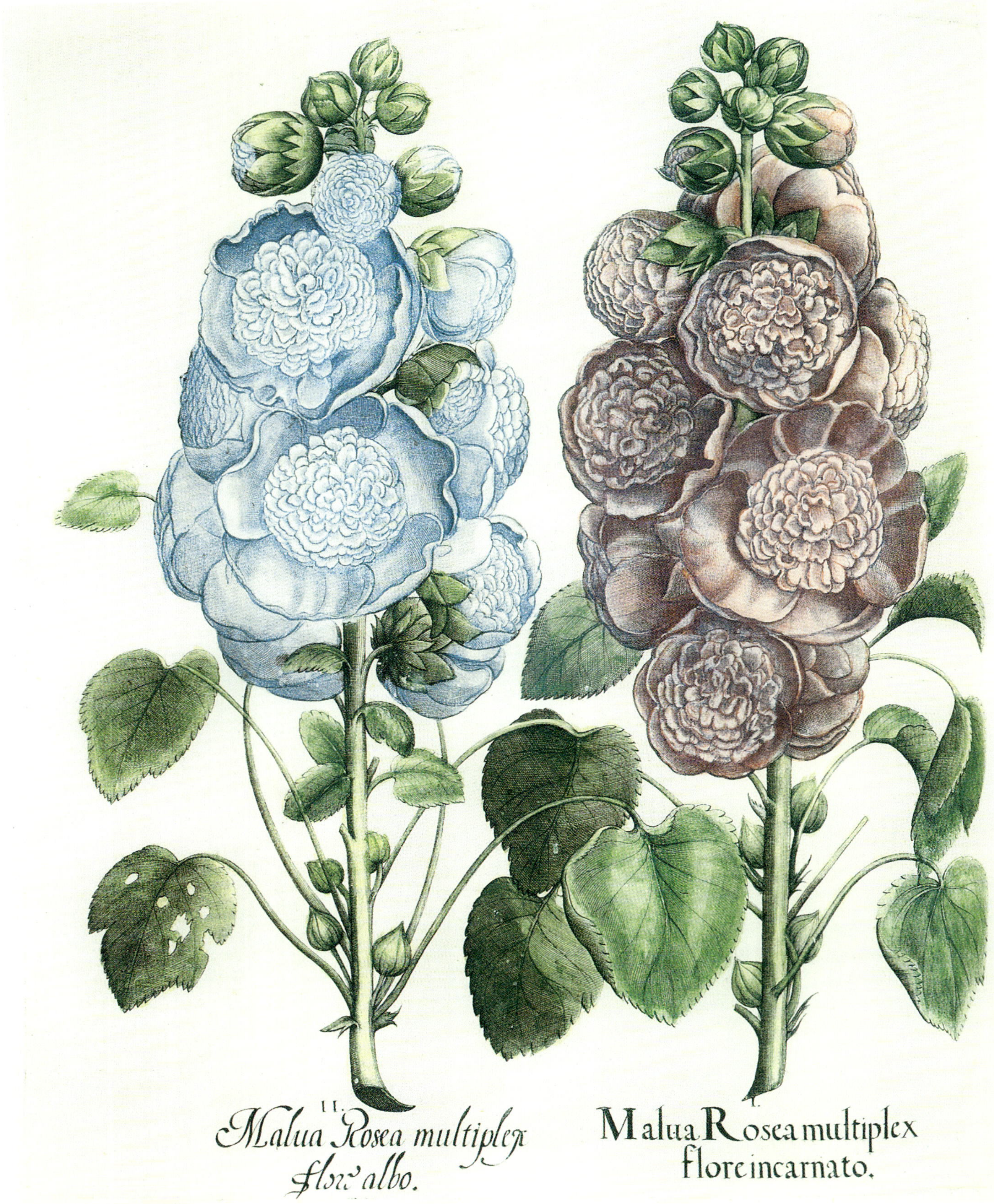

I. Double pink hollyhock *(Alcea rosea)*.
II. Double blue hollyhock *(Alcea rosea)*.

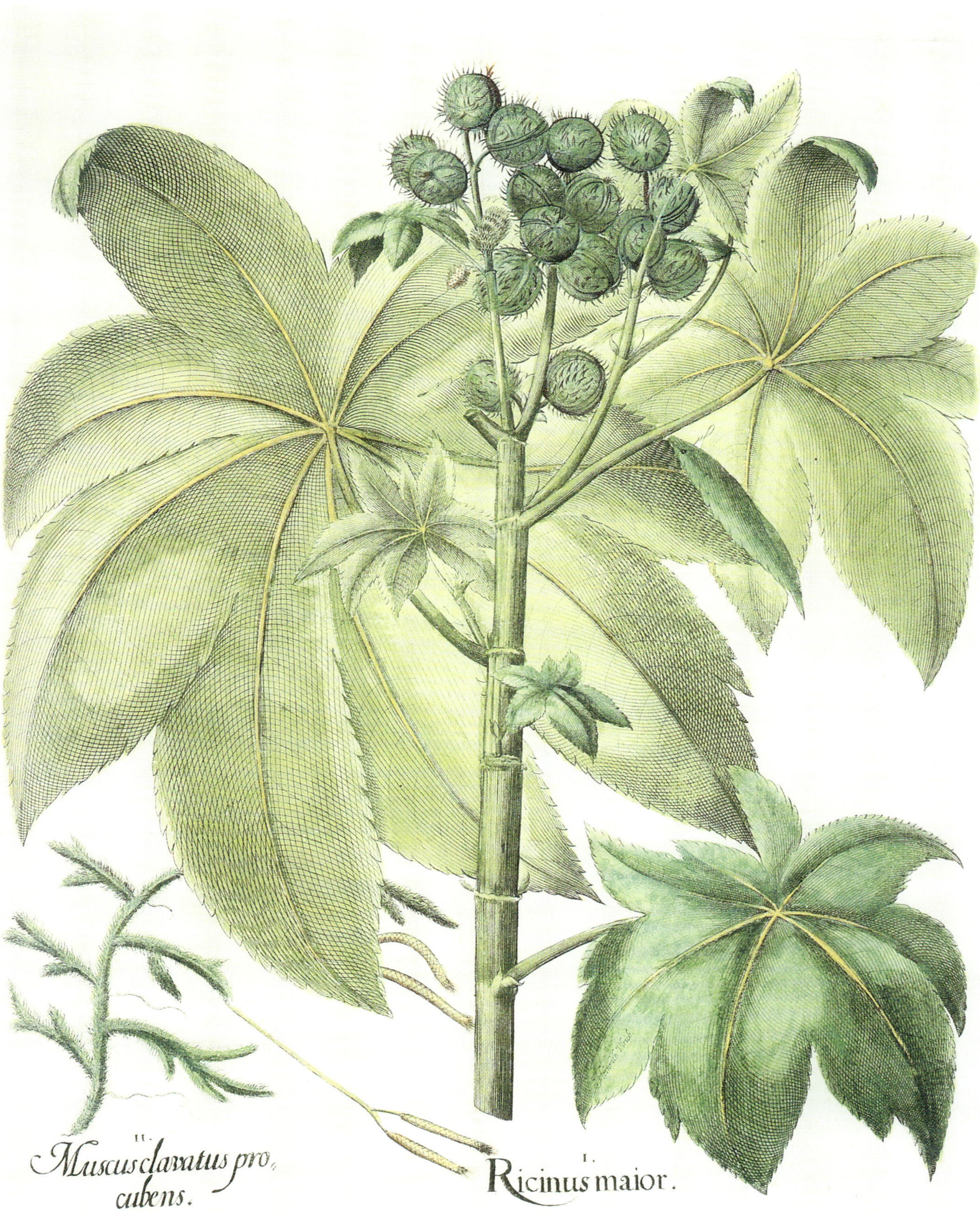

I. Castor oil plant *(Ricinus communis).*
II. Ground pine *(Lycopodium clavatum).*

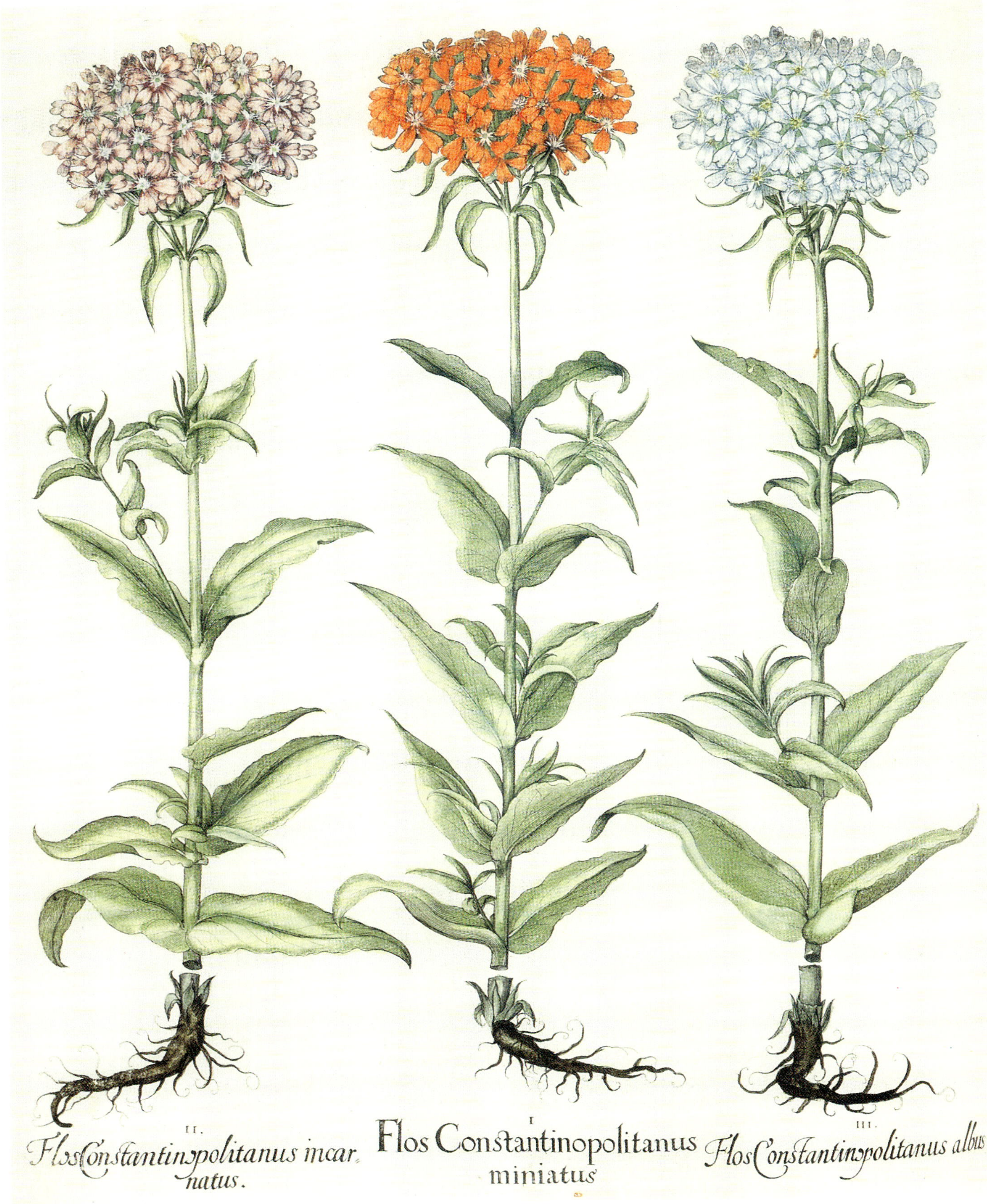

I. Maltese cross with red flowers *(Lychnis chalcedonica).* **II.** Maltese cross with pink flowers *(Lychnis chalcedonica).* **III.** Maltese cross with white flowers *(Lychnis chalcedonica).*

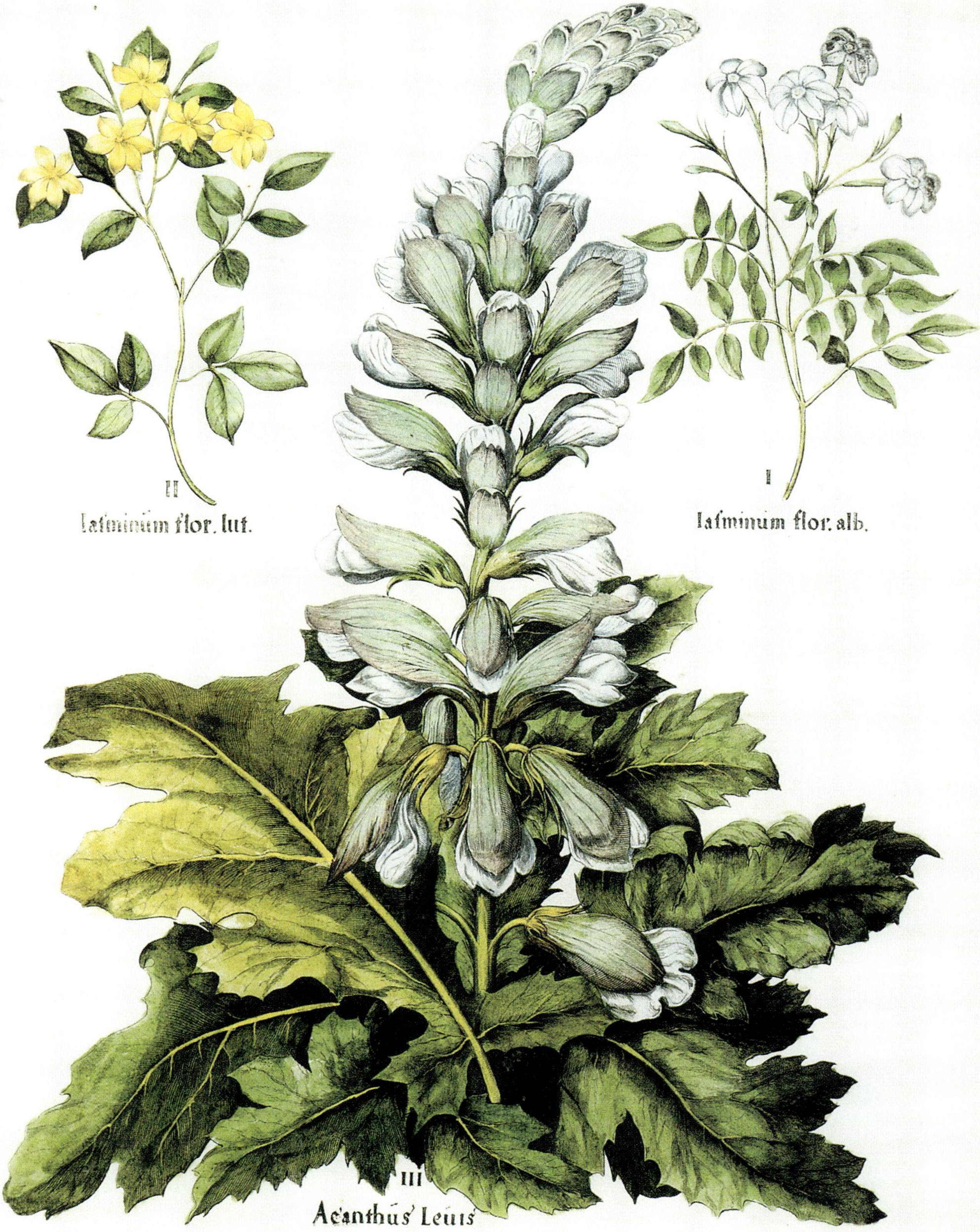

I. Common jasmine *(Jasminum grandiflorum* or *Jasminum officinale).* **II.** Jasmine *(Jasminum odoratissimum* or *Jasminum fruticans).* **III.** Bear's breeches *(Acanthus mollis).*

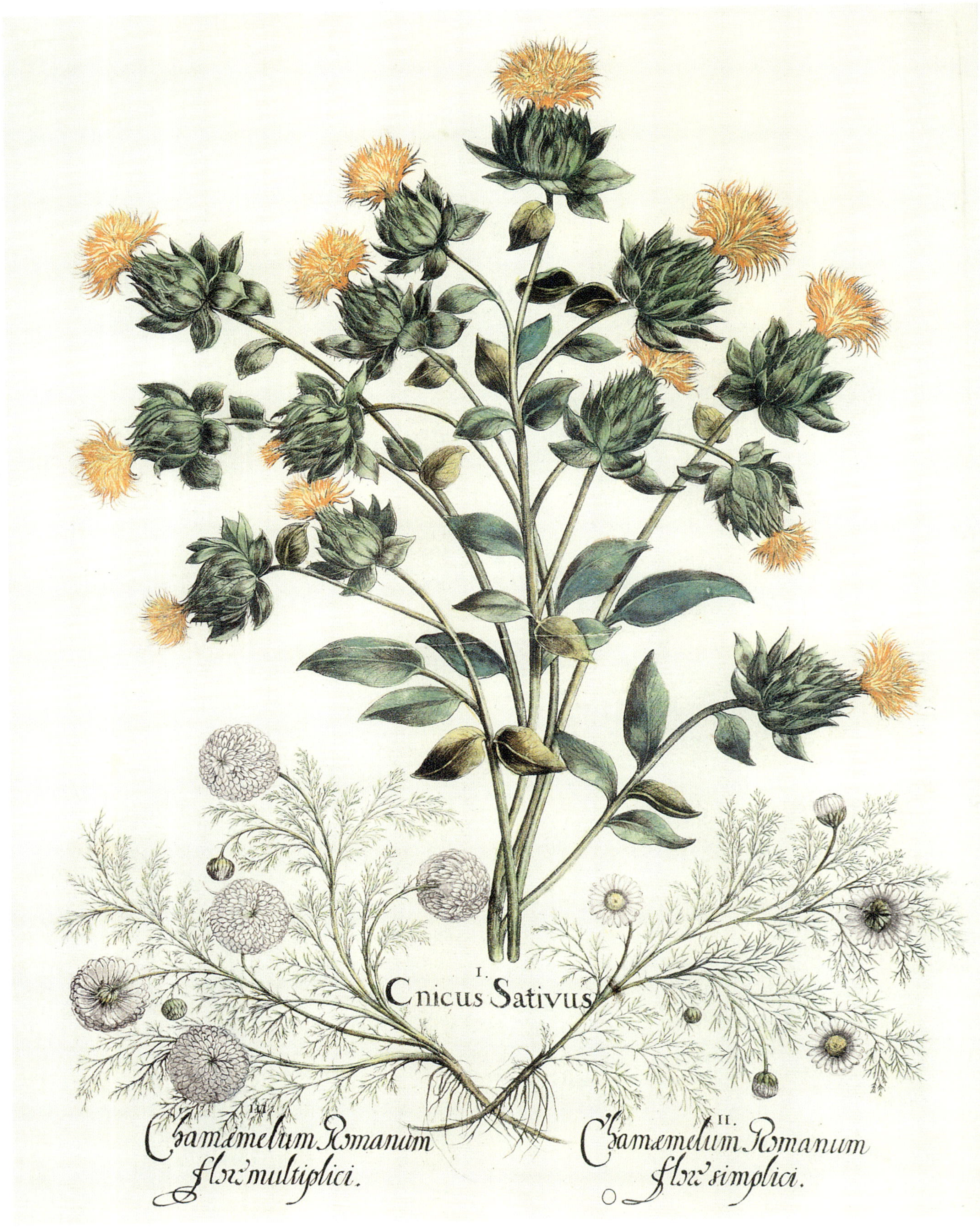

I. Safflower *(Carthamus tinctorius).* **II, III.** Chamomile *(Chamaemelum nobile).*

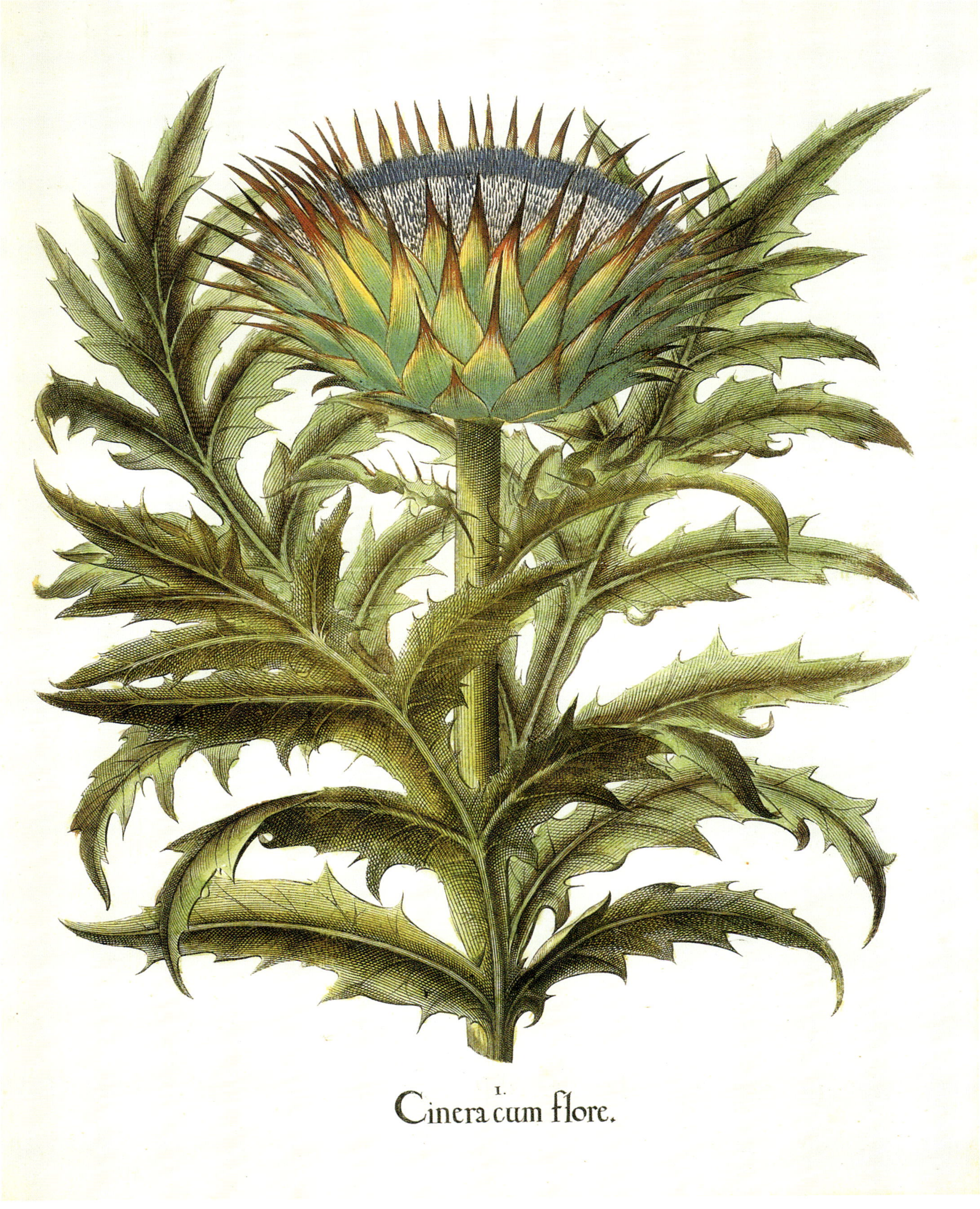

I. Cardoon
(Cynara cardunculus).

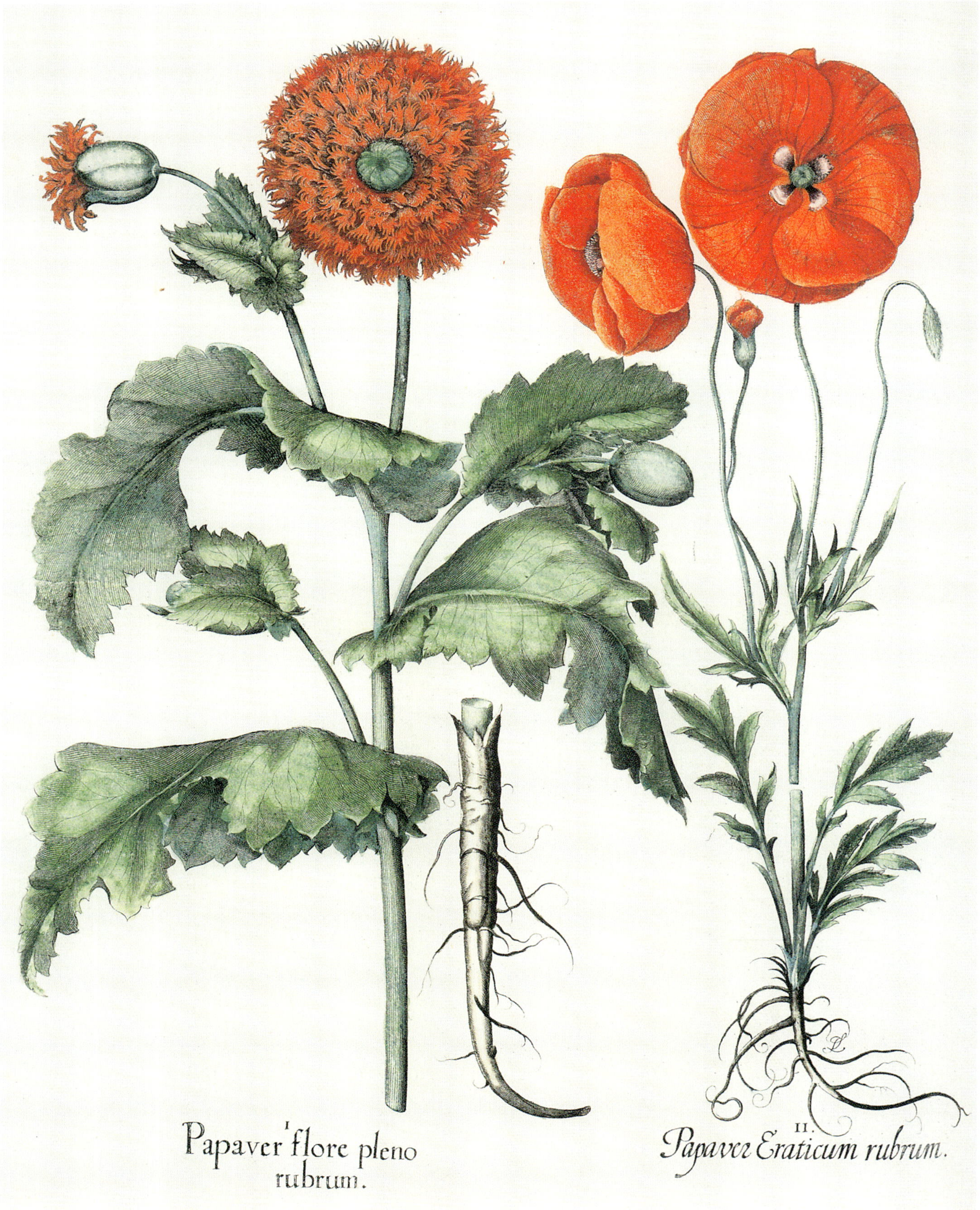

I. Double opium poppy *(Papaver somniferum).*
II. Corn poppy *(Papaver rhoeas).*

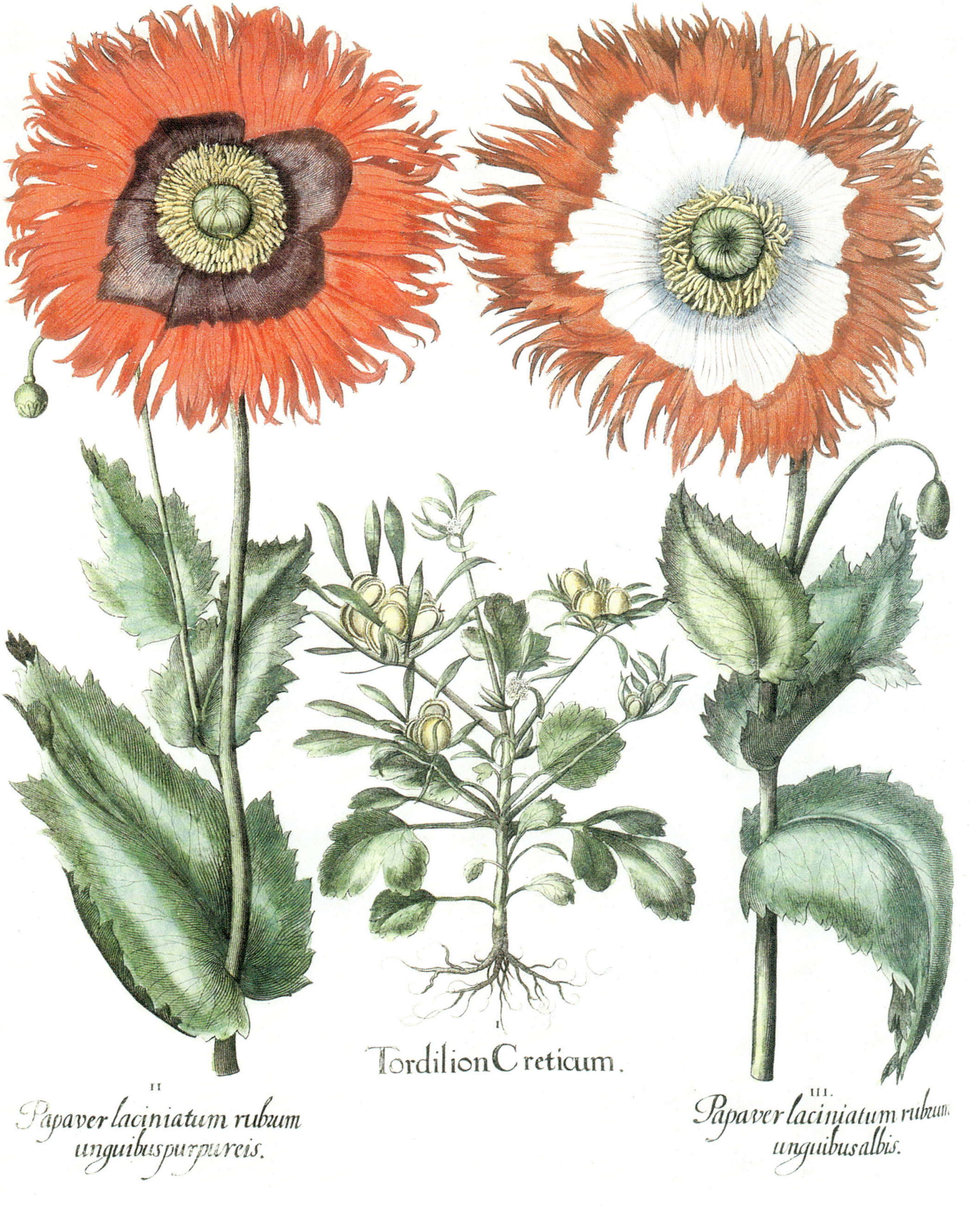

I. Hartwort *(Tordylium apulum* or *Tordylium officinale).* **II, III.** Double opium poppy with bi-colored flowers *(Papaver somniferum).*

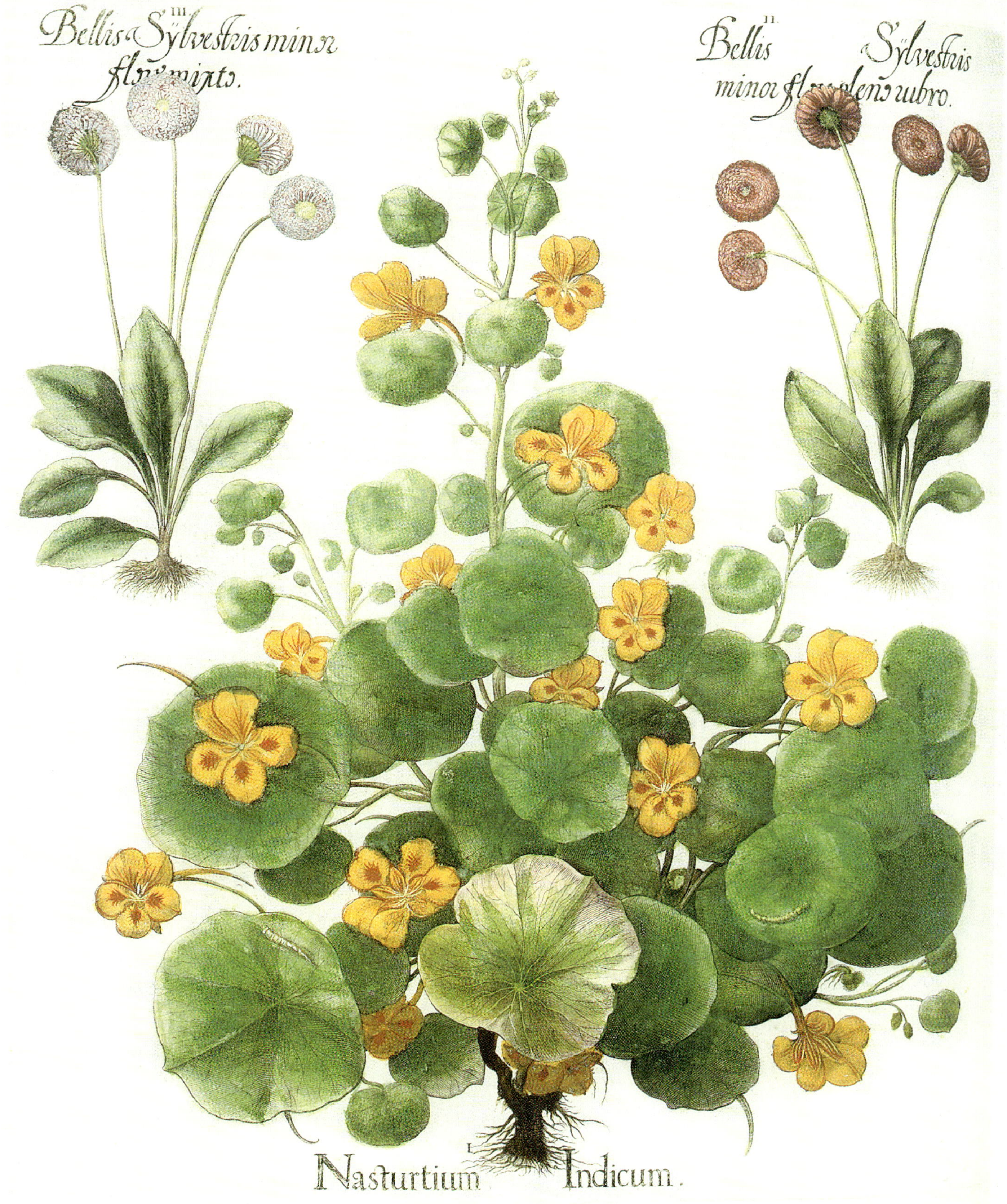

I. Nasturtium *(Tropaeolum minus)*.
II, III. Daisy *(Bellis spec.)*.

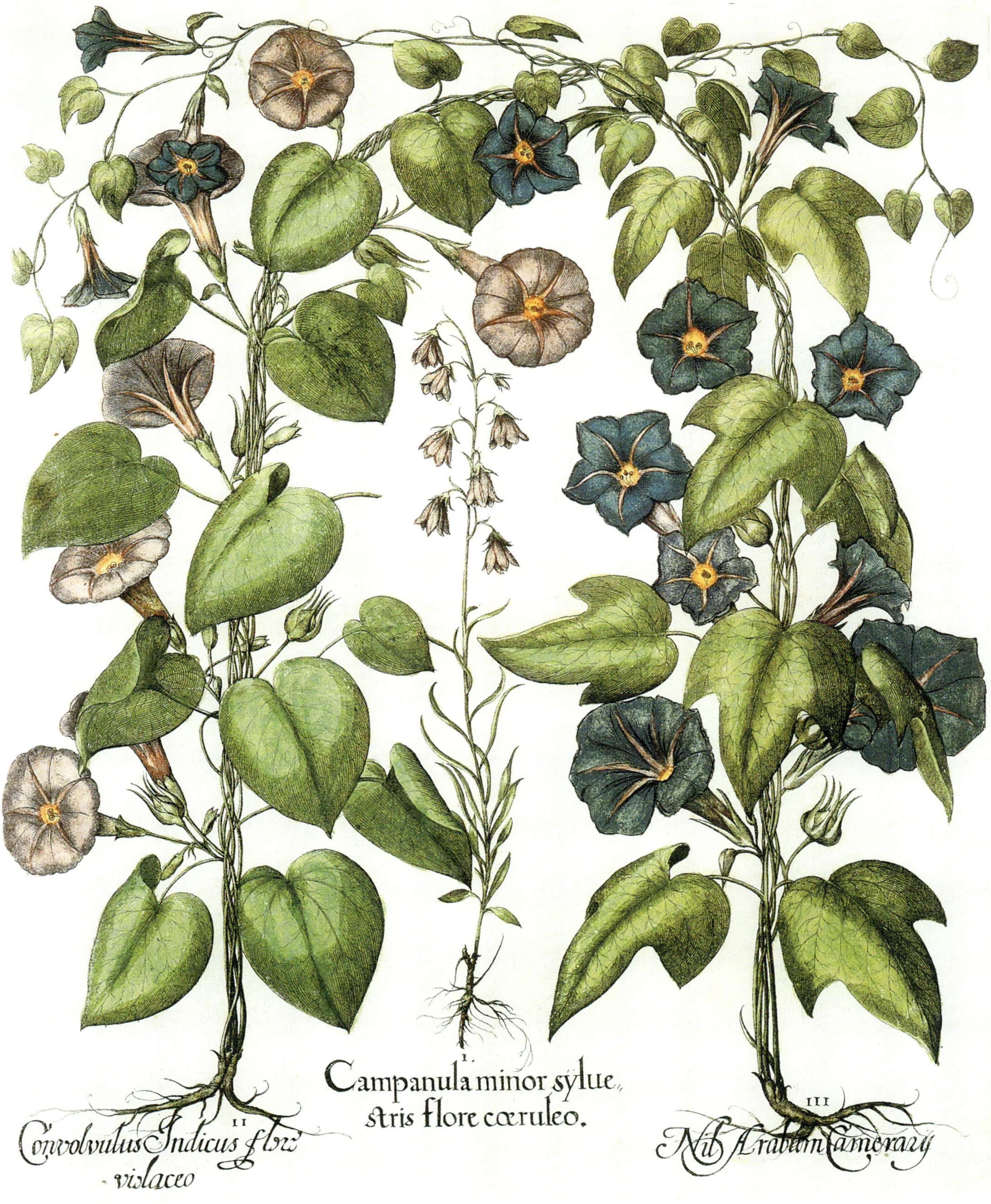

I. Bluebell *(Campanula rotundifolia).* **II.** Common morning glory *(Pharbitis purpurea).* **III.** Morning glory *(Pharbitis nil).*

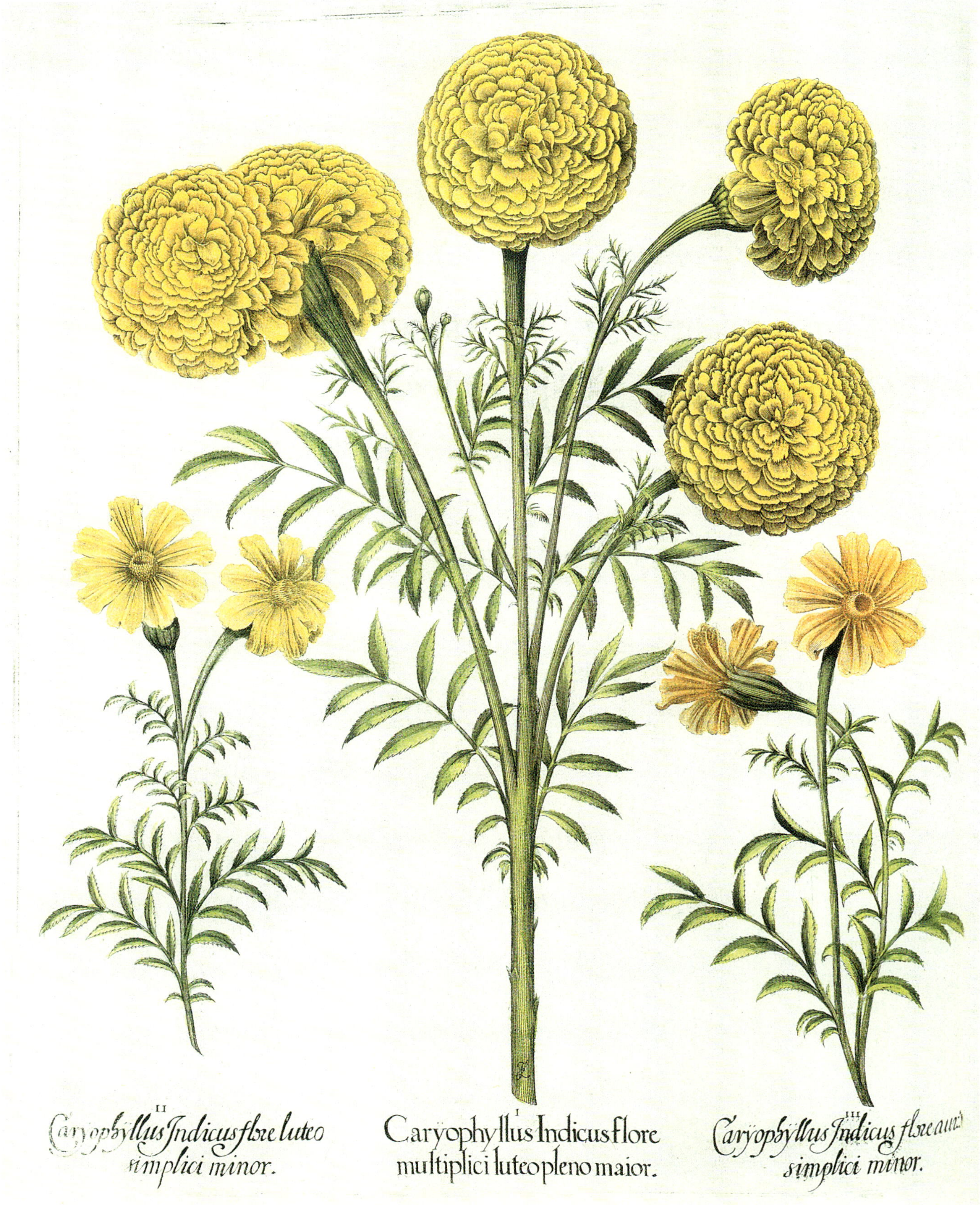

I. African marigold *(Tagetes erecta)*. **II, III.** French marigold *(Tagetes patula)*.

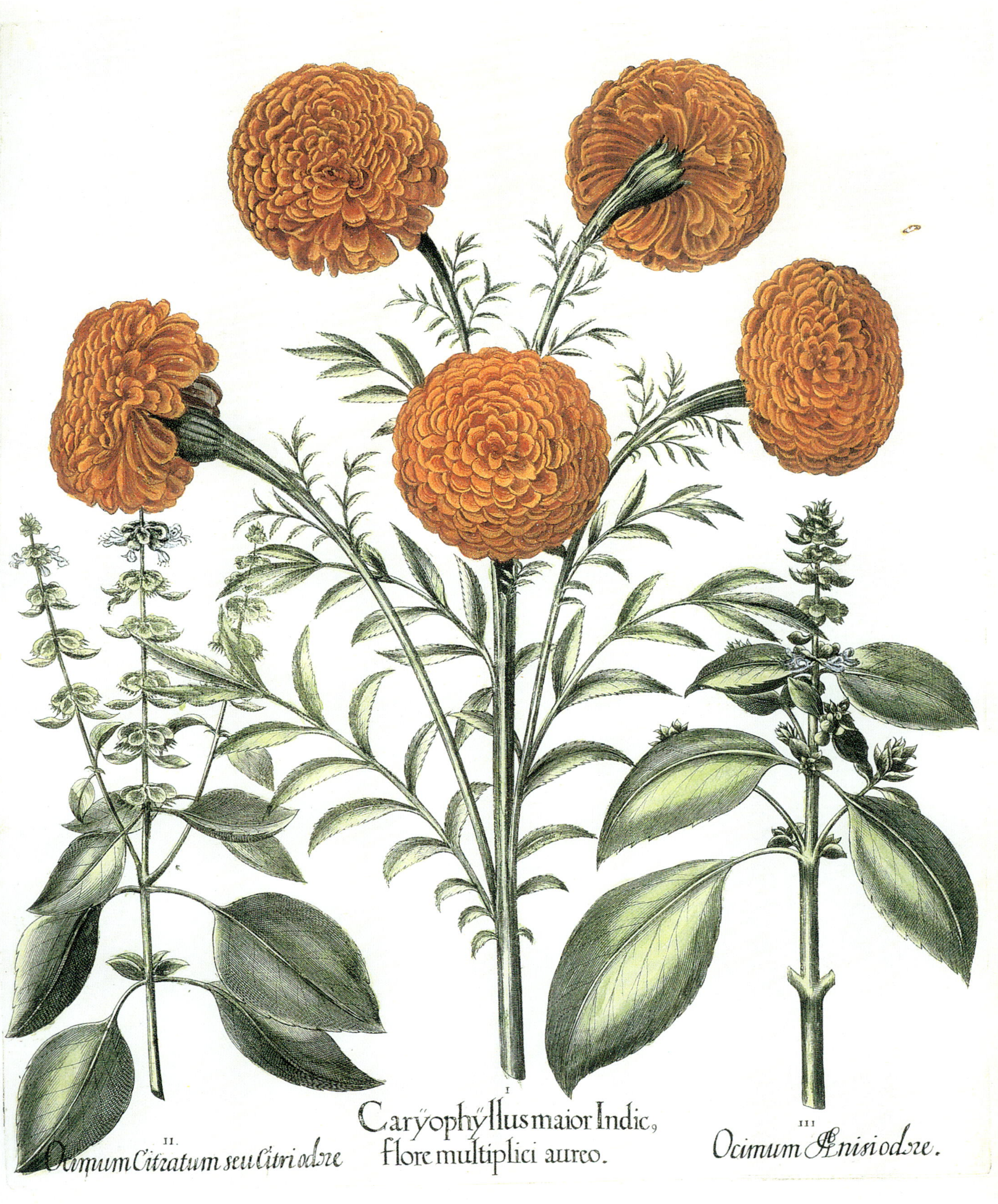

I. African marigold *(Tagetes erecta).* **II, III.** Common basil *(Ocimum basilicum).*

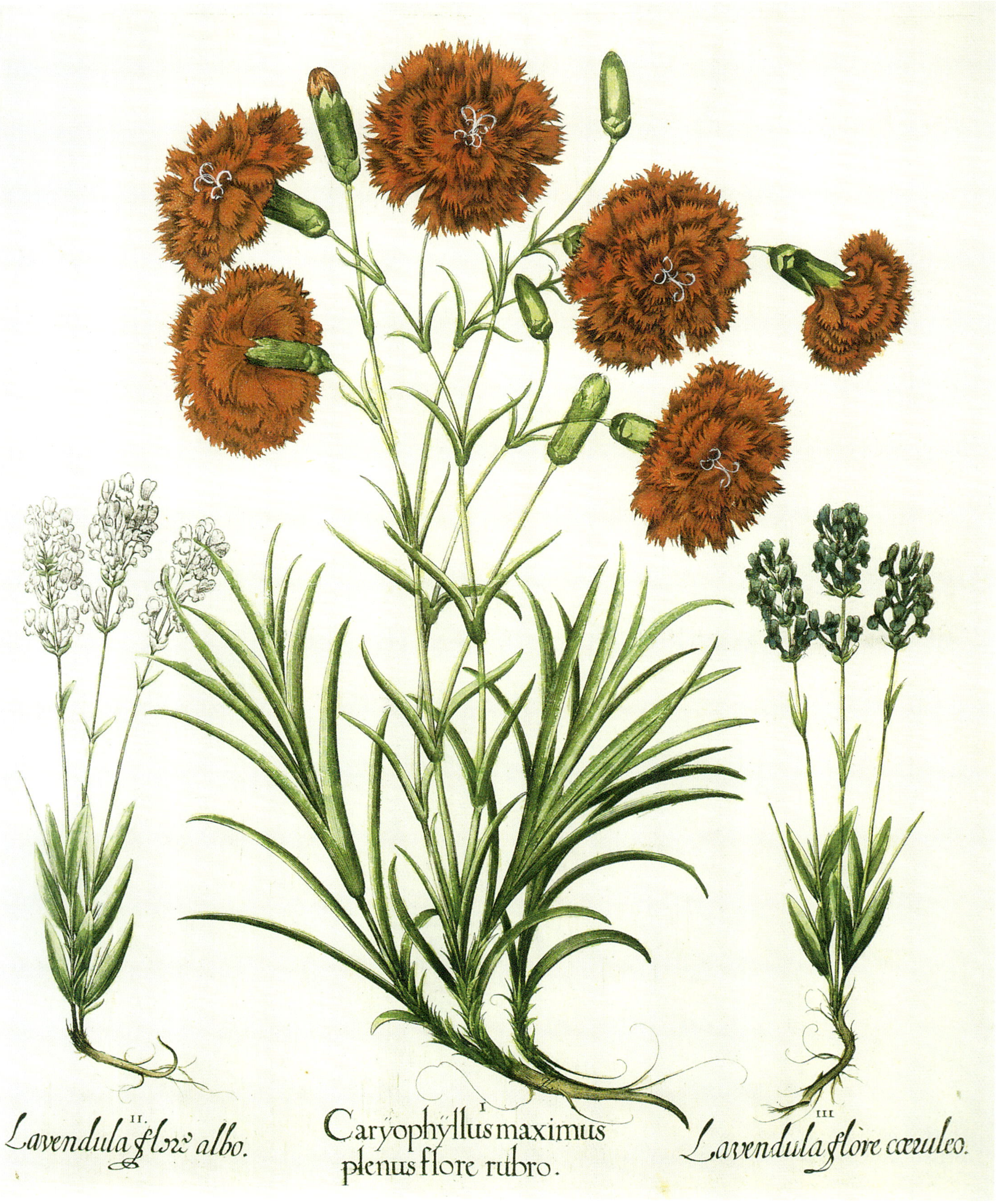

I. Wild carnation *(Dianthus caryophyllus).*
II, III. Lavender *(Lavandula spec.).*

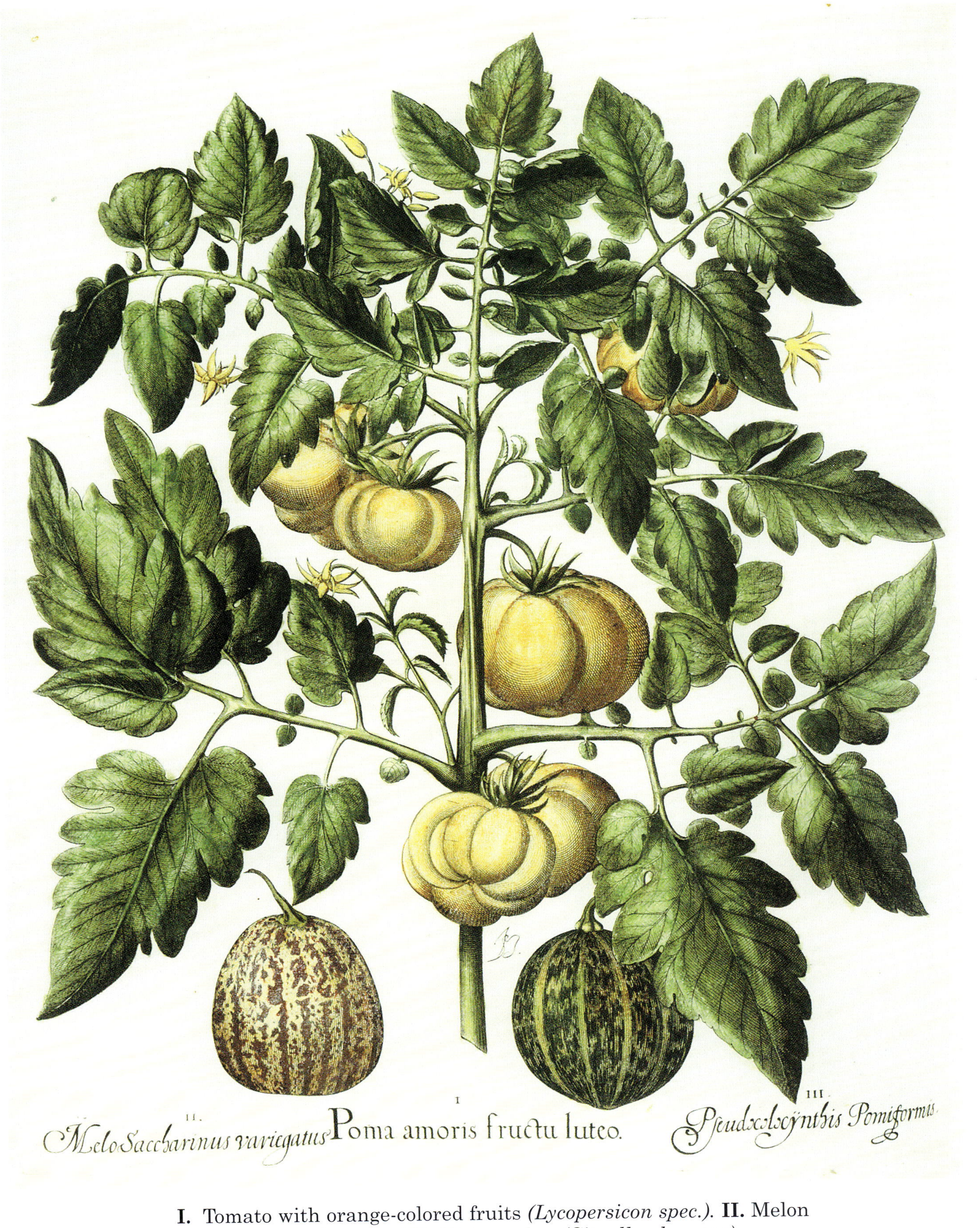

I. Tomato with orange-colored fruits *(Lycopersicon spec.)*. **II.** Melon *(Cucumis melo)*. **III.** Watermelon *(Citrullus lanatus)*.

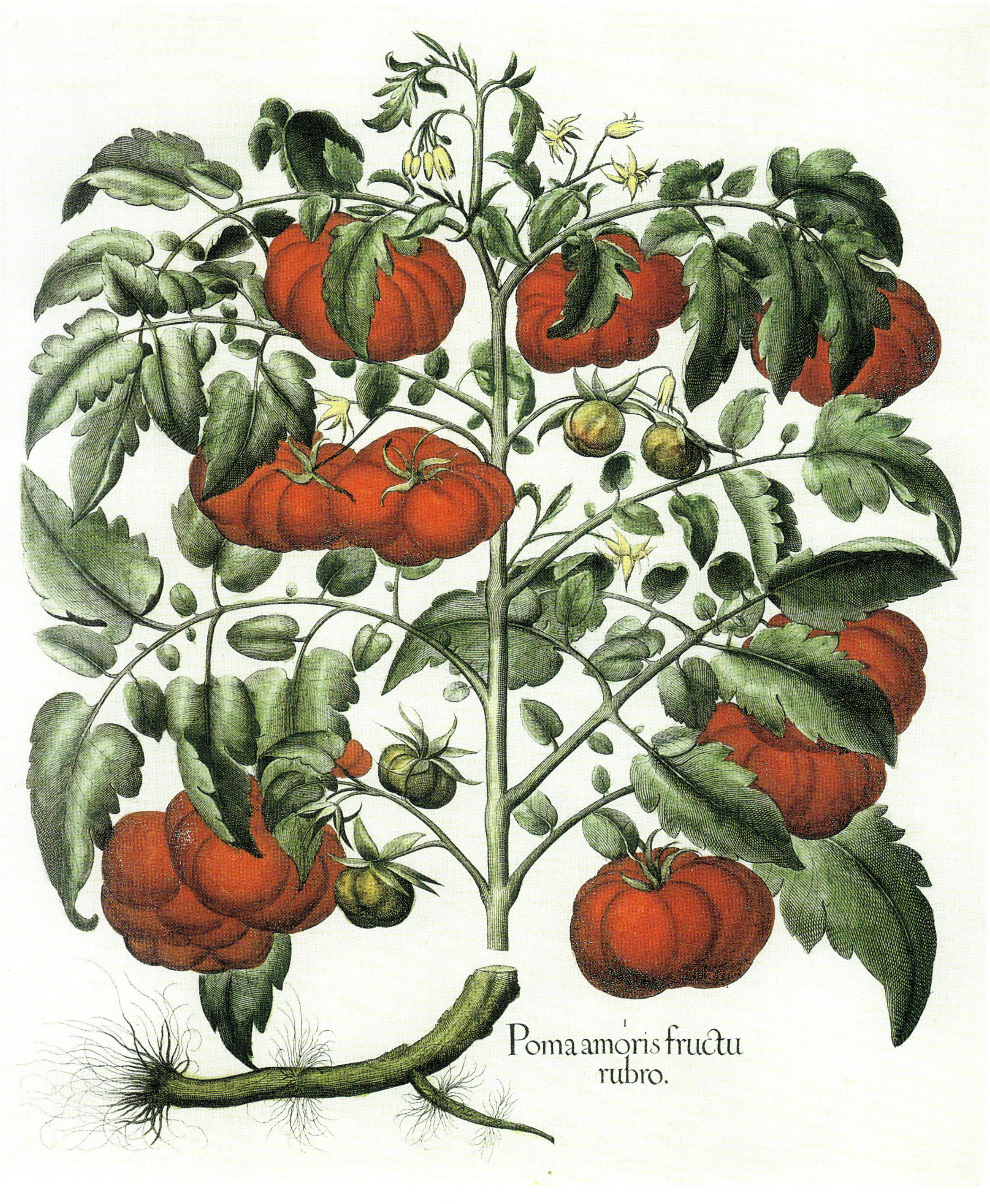

I. Tomato
(Lycopersicon esculentum).

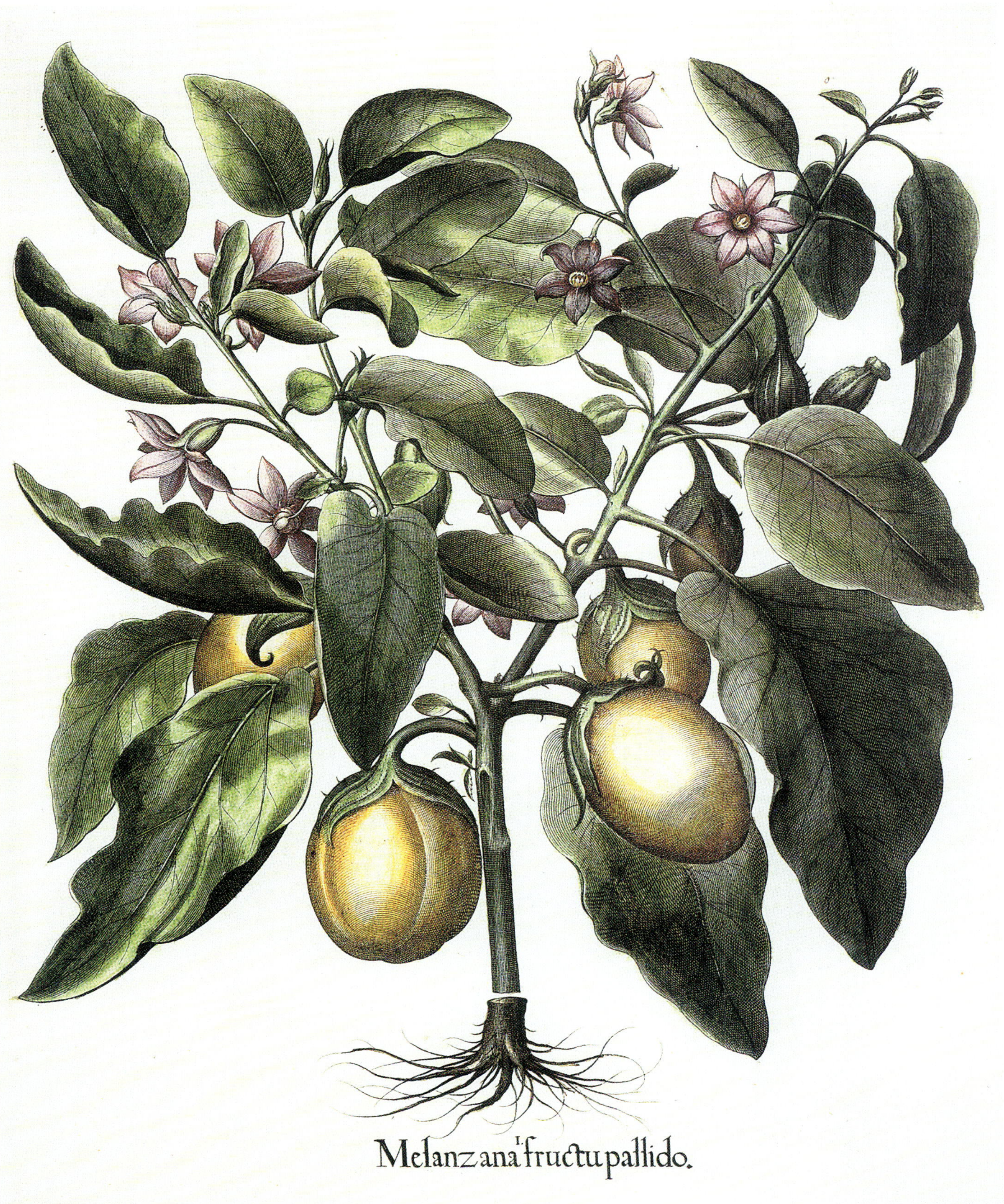

I. Eggplant
(Solanum melongena).

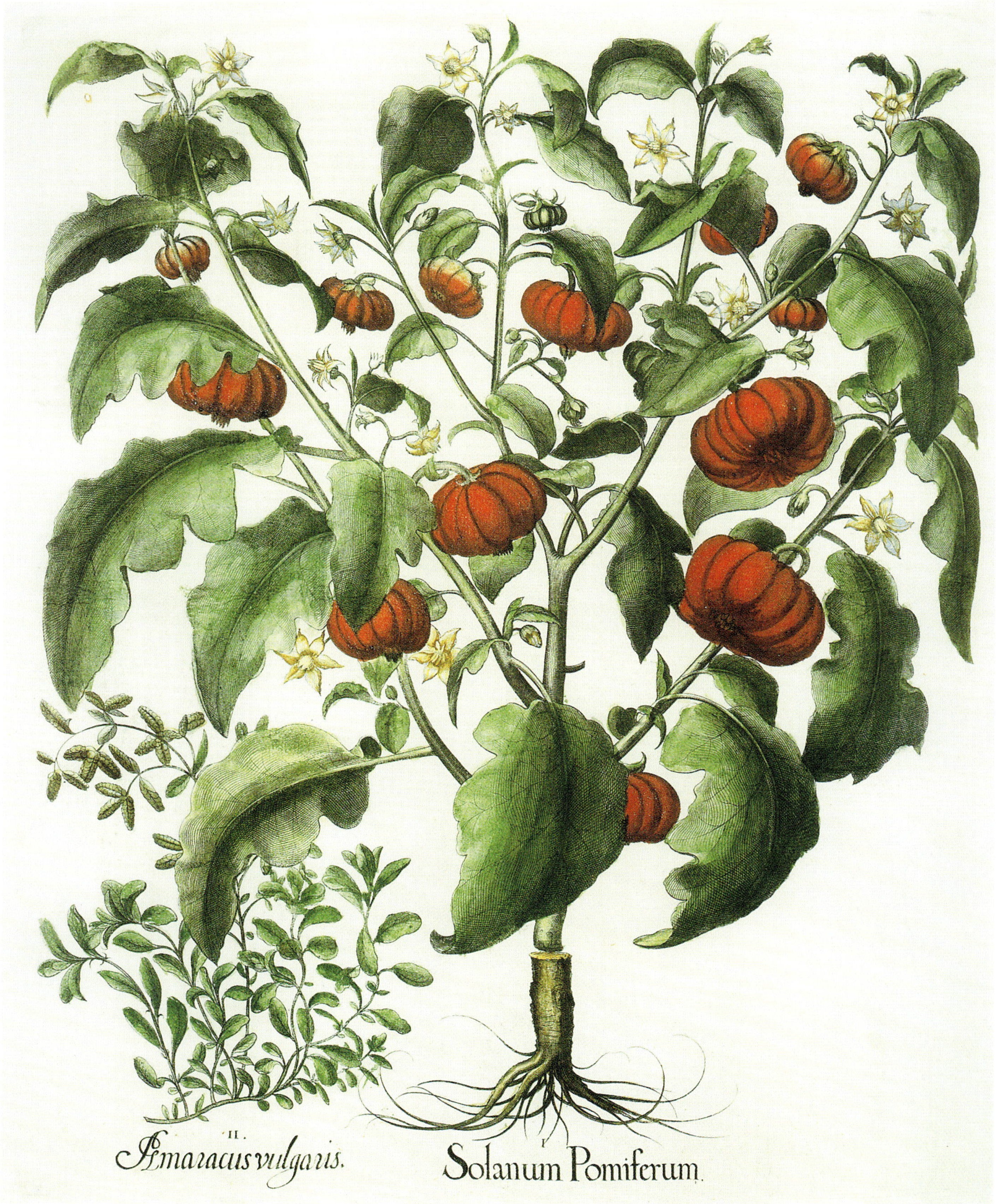

I. Nightshade *(Solanum macrocarpon).*
II. Marjoram *(Origanum majorana).*

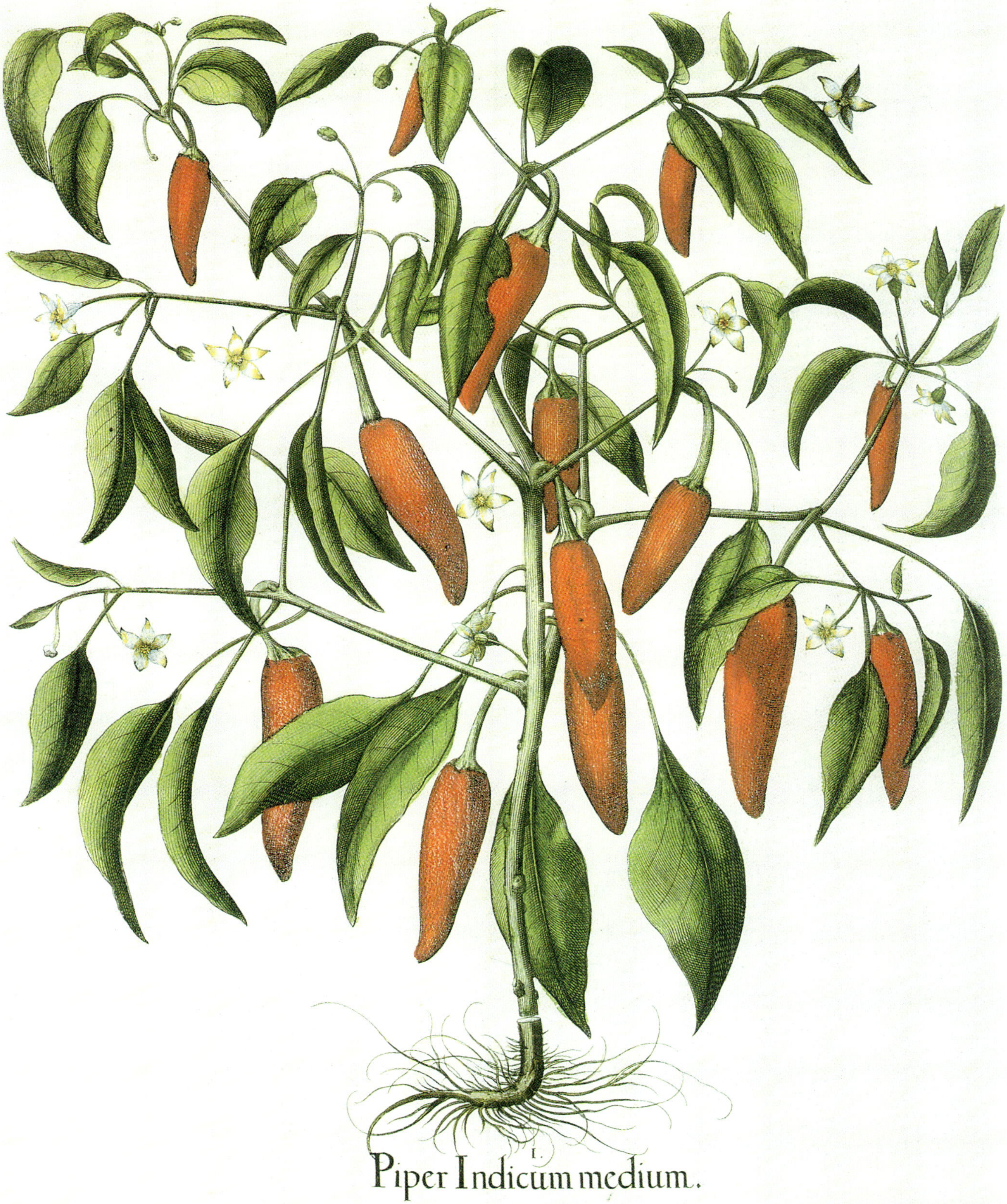

I. Chili pepper
(Capsicum spec.).

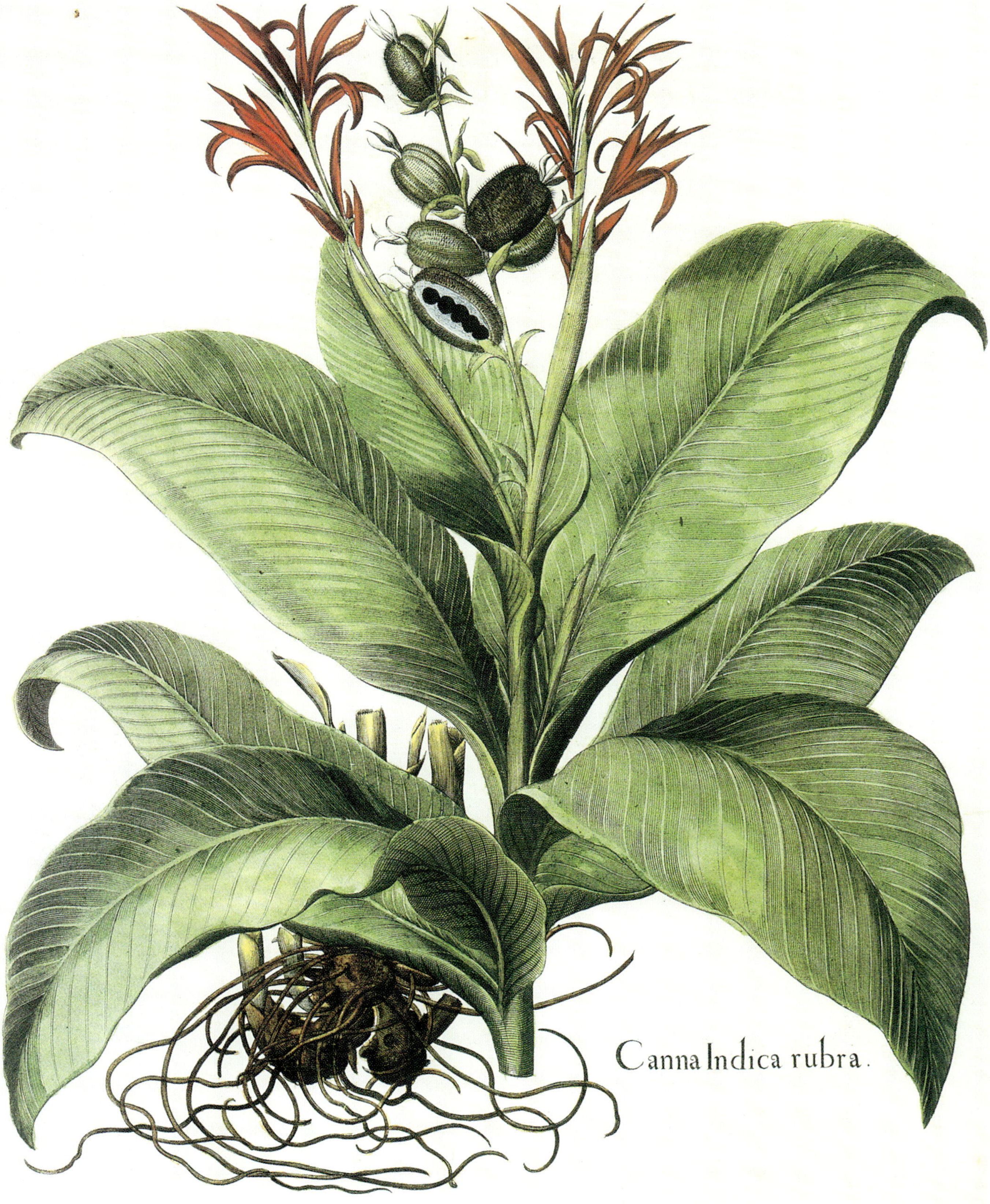

I. Red Indian shot
(Canna indica).

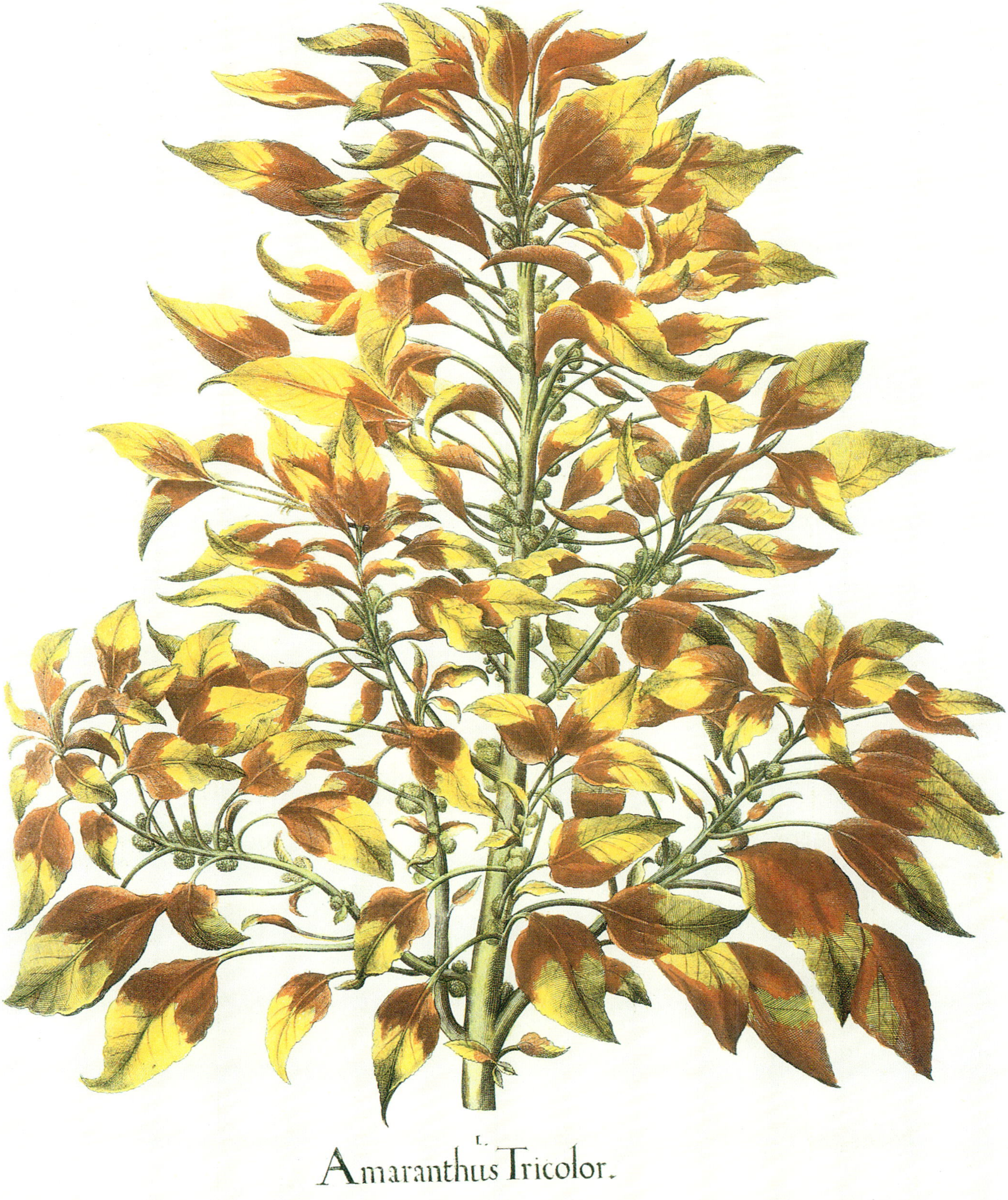

I. Tampala
(Amaranthus tricolor).

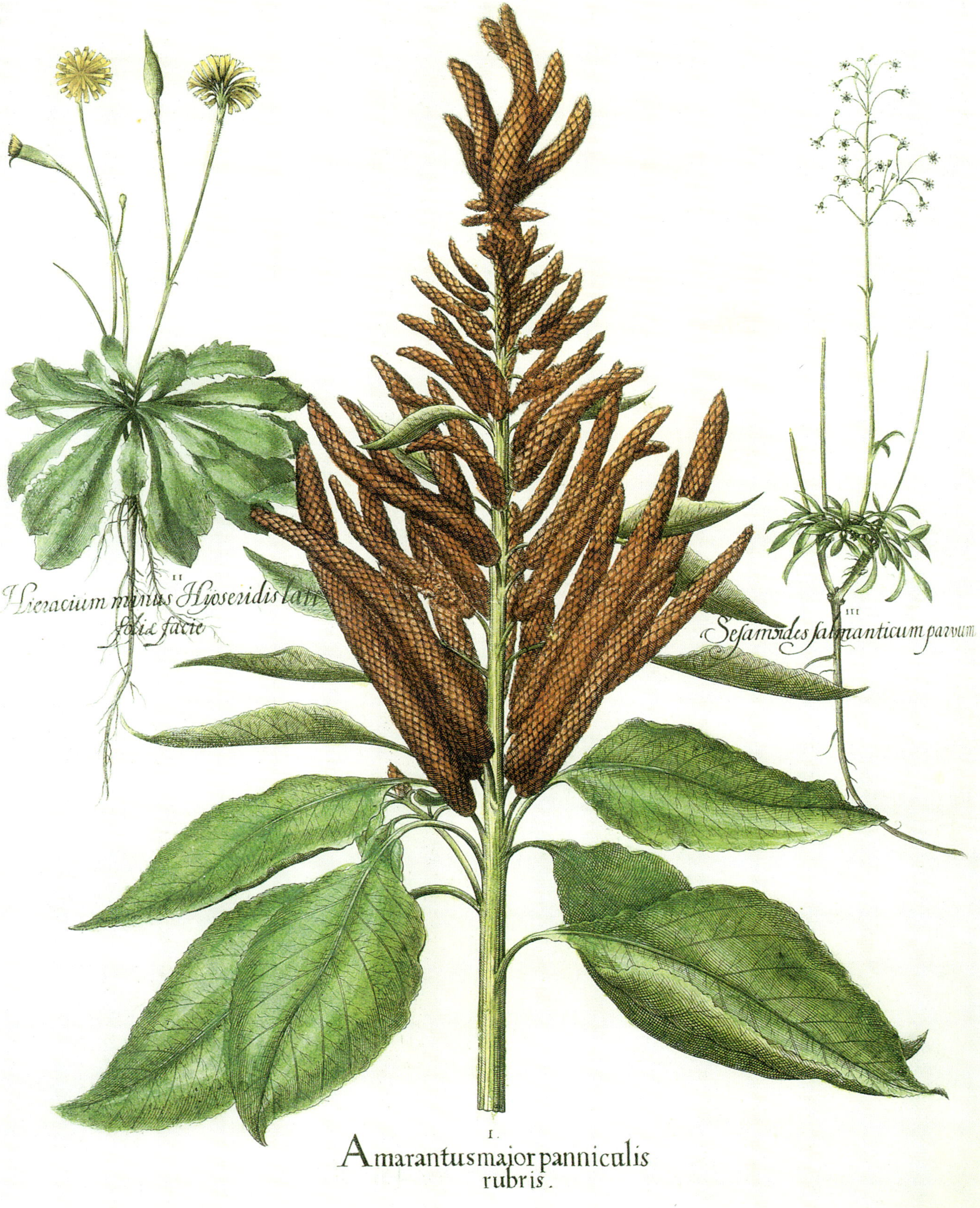

I. Purple amaranth *(Amaranthus paniculatus).* **II.** Hawk's-beard *(Crepis tectorum).*
III. Spanish catchfly *(Silene otites).*

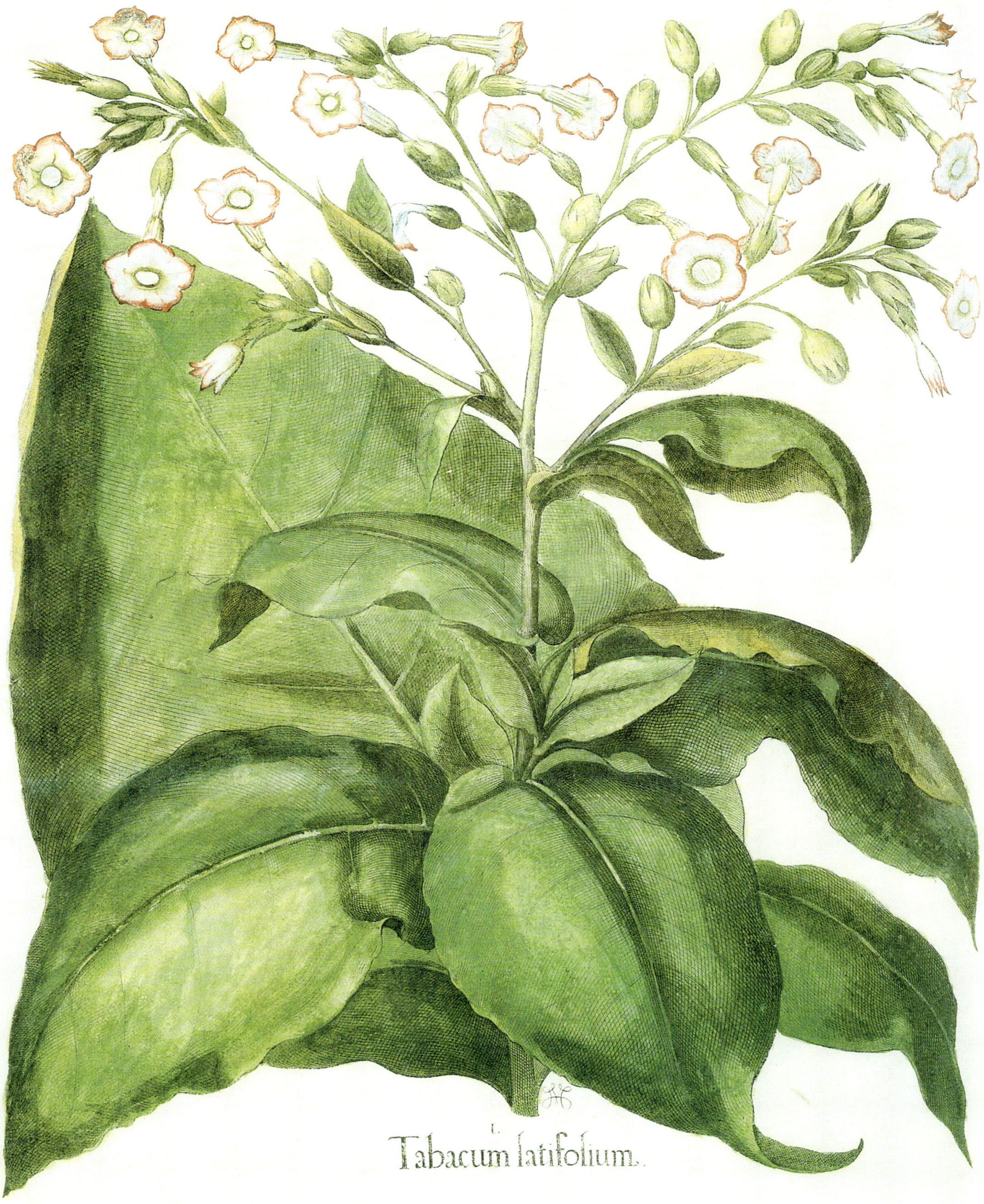

I. Tobacco
(Nicotiana latissima).

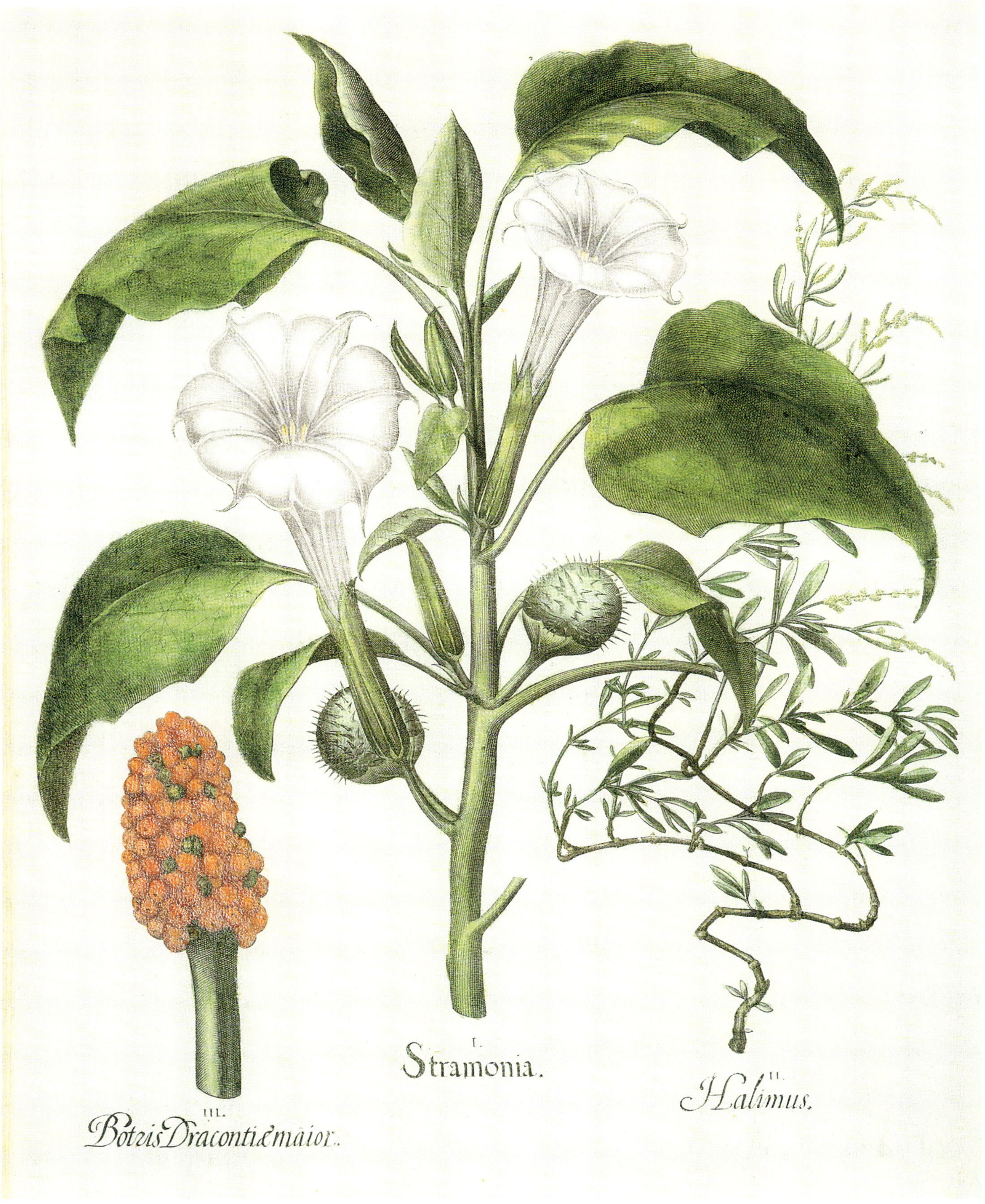

I. Horn of plenty *(Datura metel).* **II.** Sea purslane *(Halimione portulacoides).*
III. Lords-and-ladies *(Arum maculatum).*

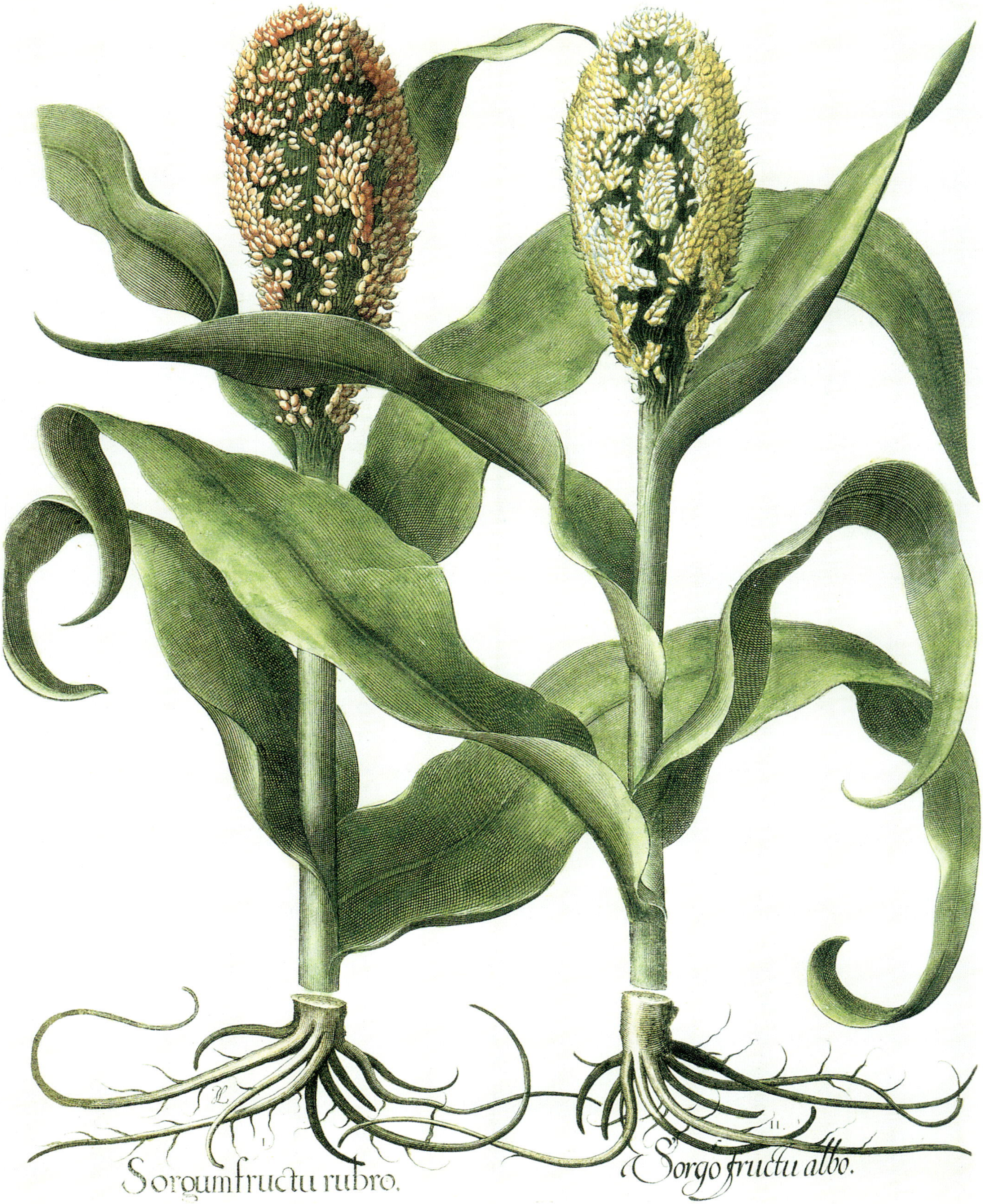

I, II. Sorghum
(Sorghum bicolor).

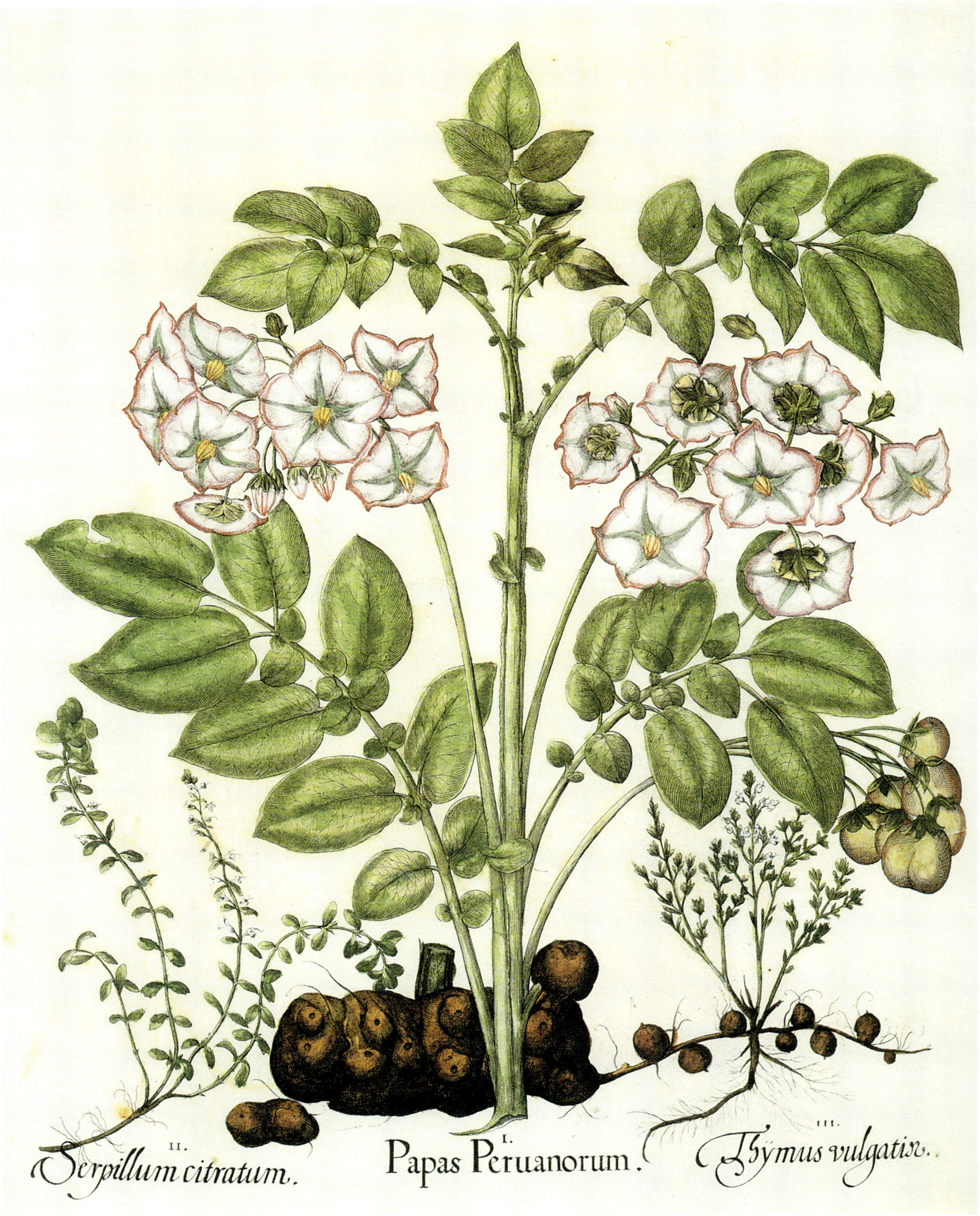

I. Potato *(Solanum tuberosum)*. **II.** Broad-leaved thyme *(Thymus pulegioides)*.
III. Common thyme *(Thymus vulgaris)*.

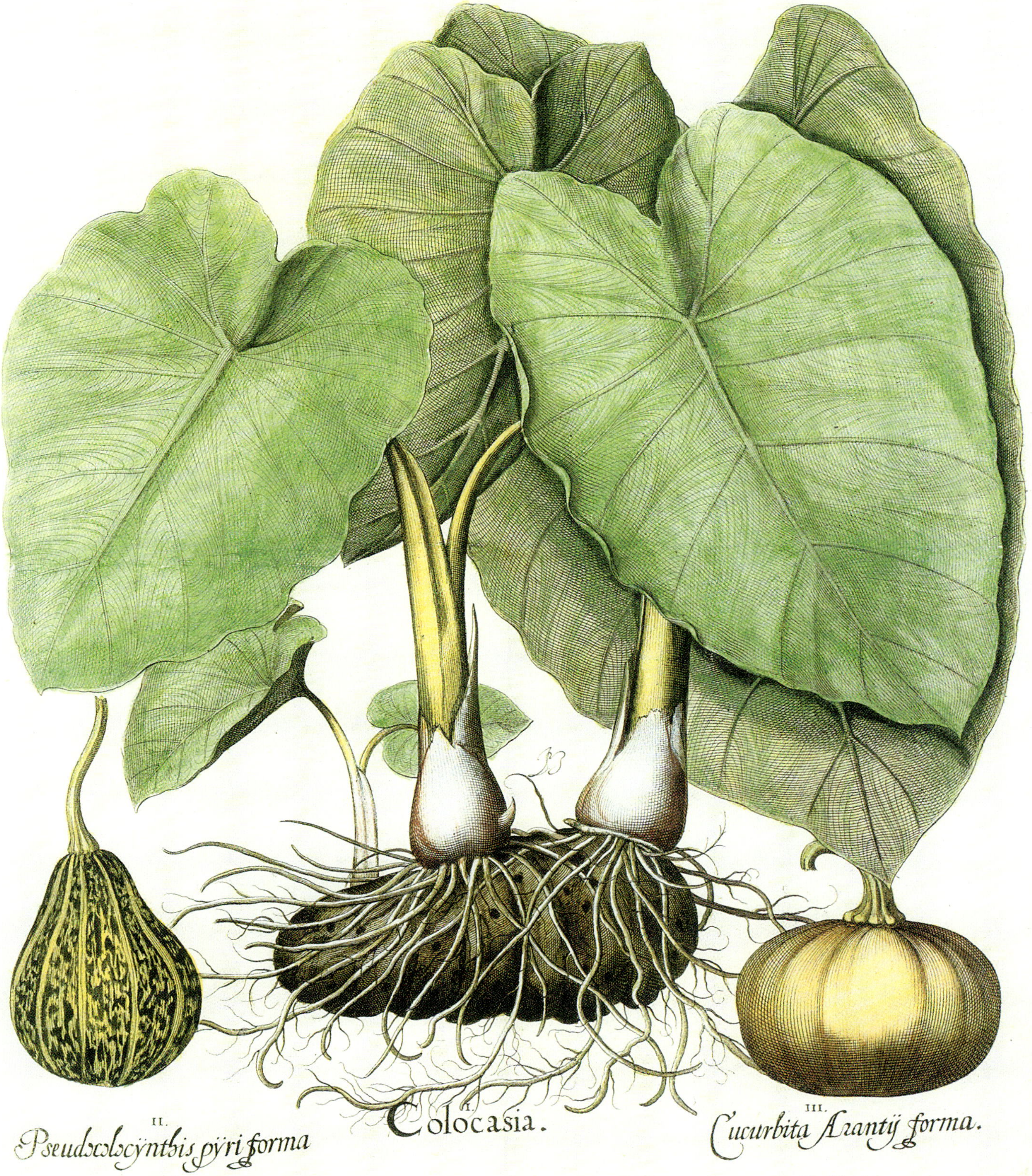

I. Cocoyam *(Colocasia esculenta).* **II.** White-flowered gourd *(Lagenaria spec.).*
III. Pumpkin *(Cucurbita pepo).*

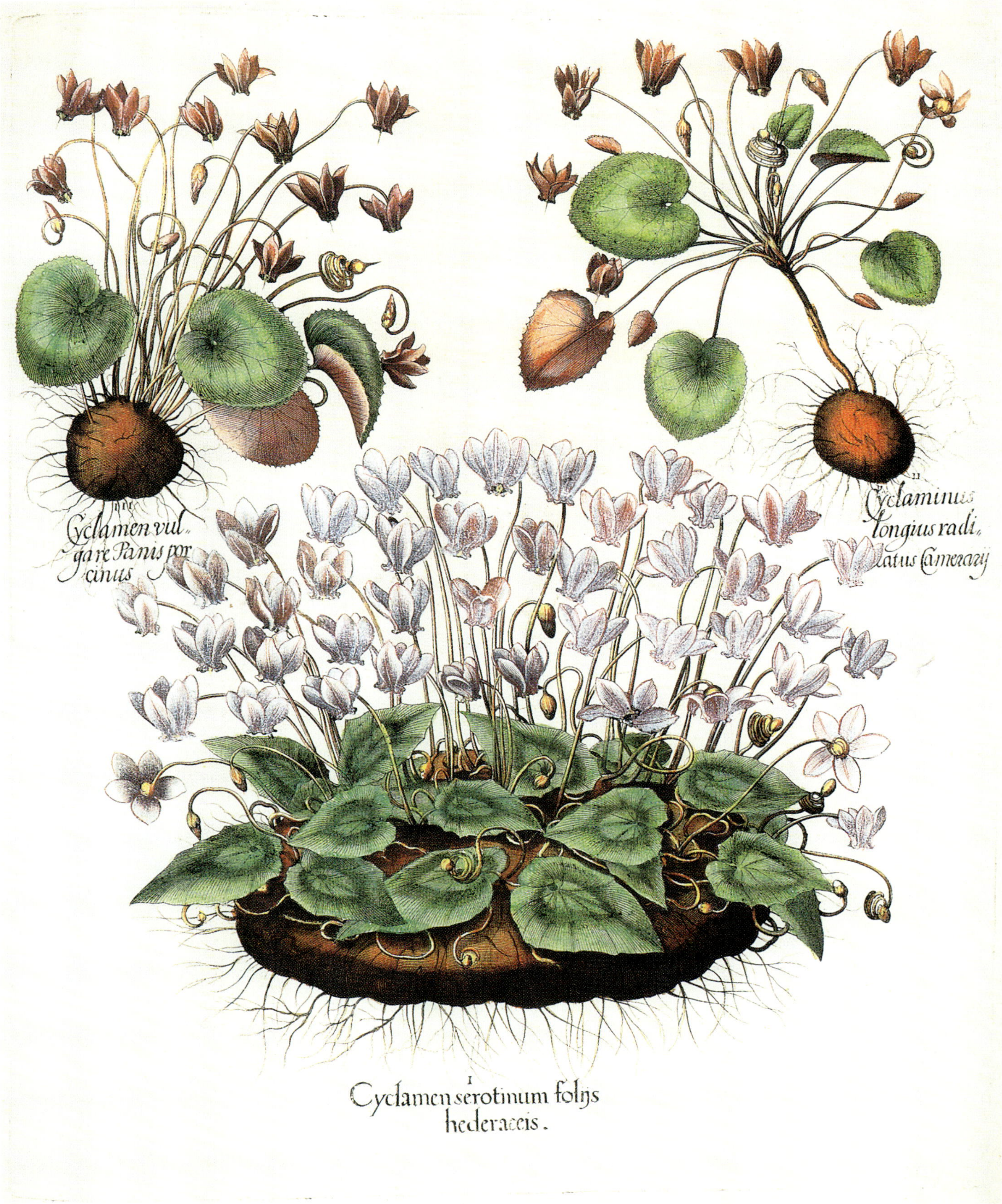

I. Sowbread *(Cyclamen hederifolium).*
II, III. Sowbread *(Cyclamen purpurascens).*

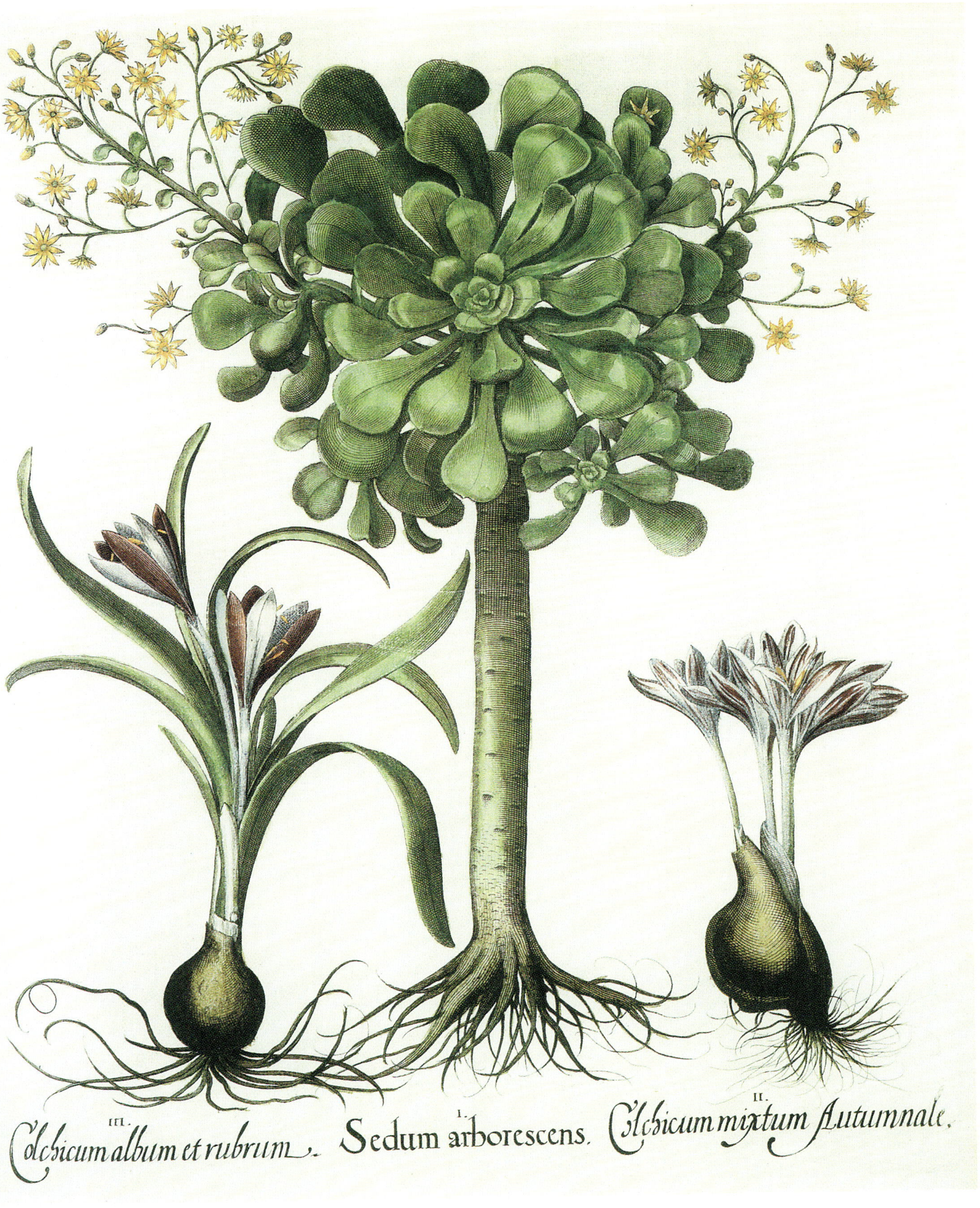

I. Houseleek *(Aeonium arboreaum).* **II.** Autumn crocus *(Colchicum spec.).*
III. Bi-colored autumn crocus in leaf *(Colchicum spec.).*

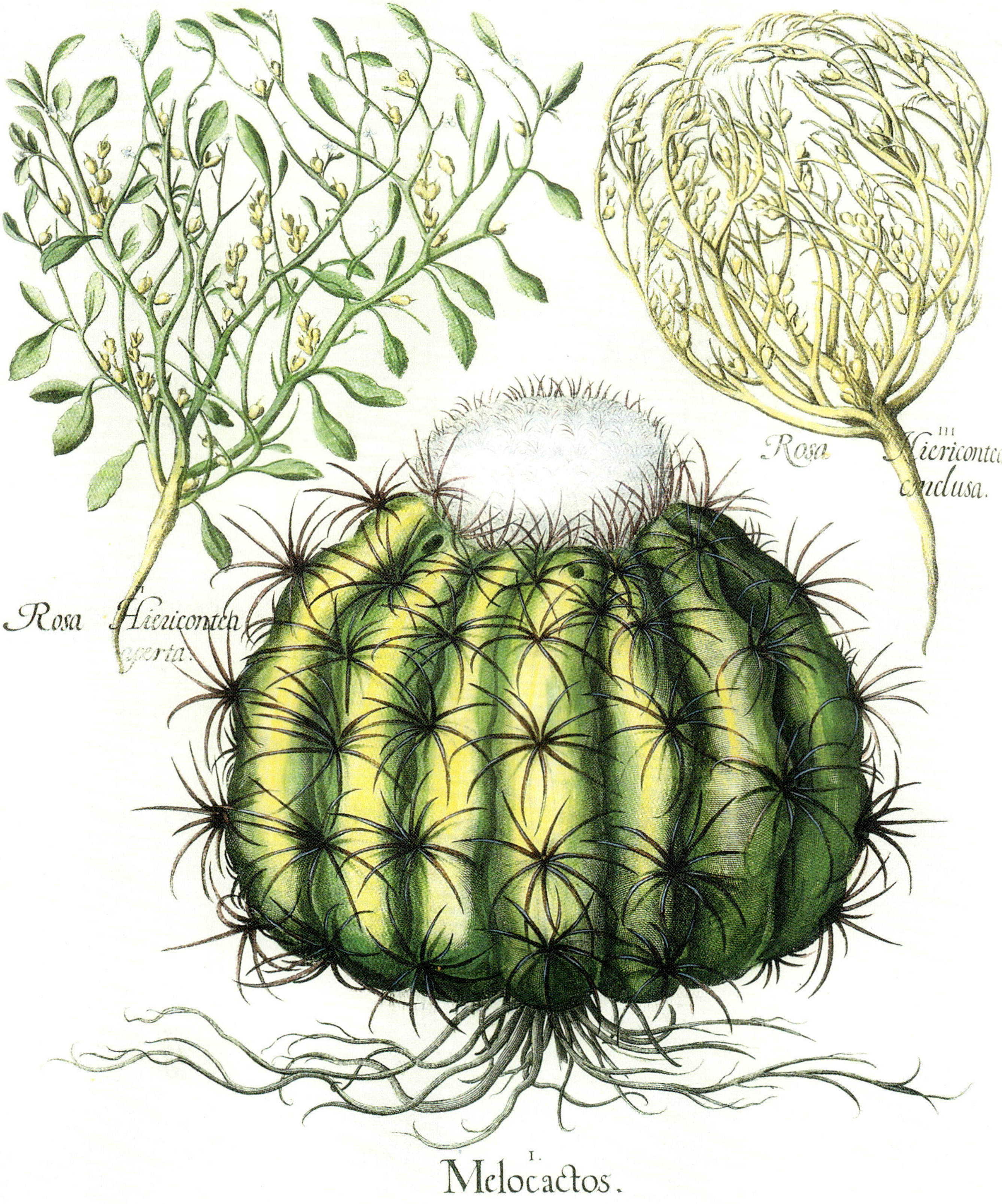

I. Marvel of Peru *(Mirabilis jalapa).* **II.** Fringed gentian *(Gentianella ciliata).*
III. Chiltern gentian *(Gentiana germanica).*

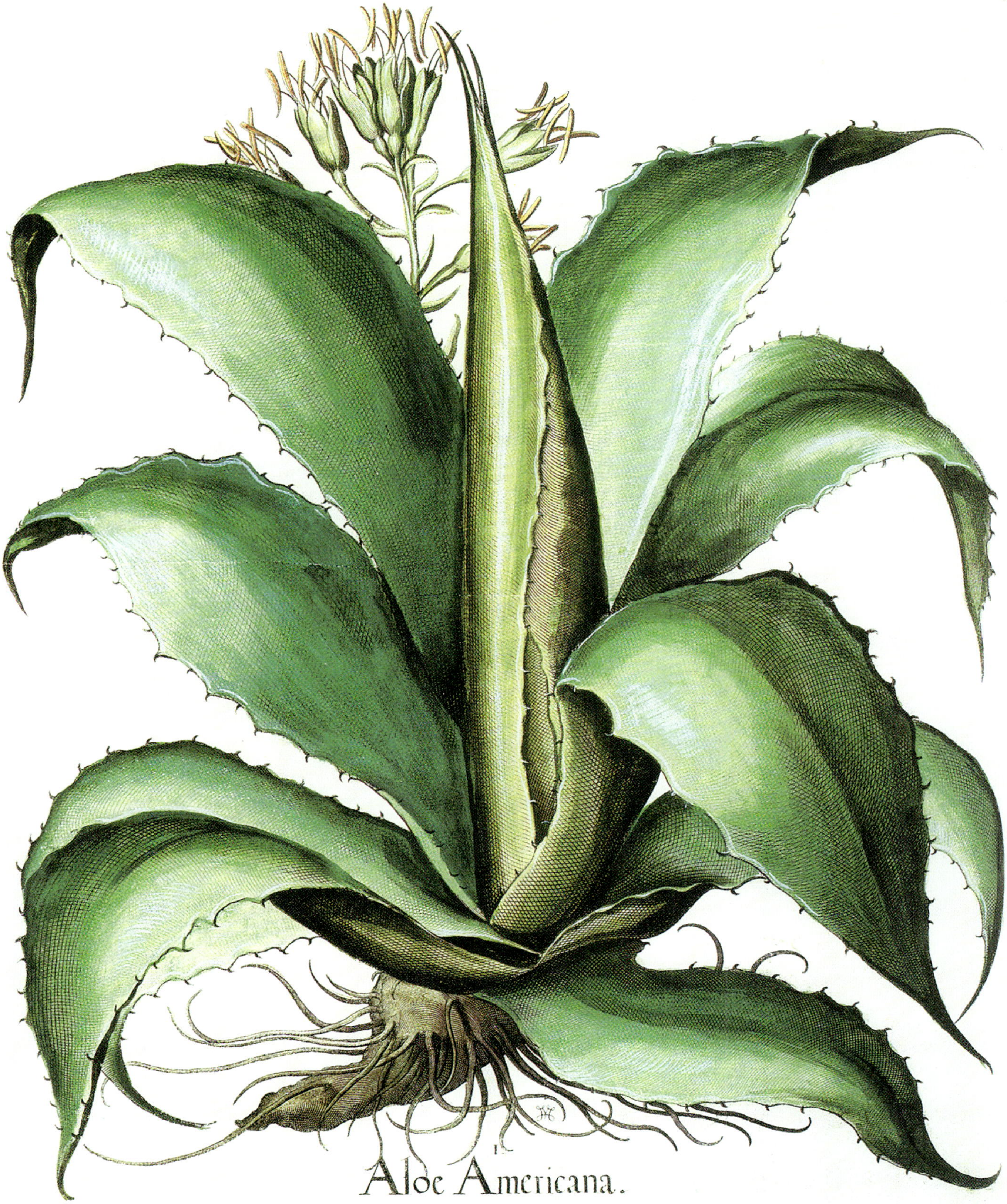

I. Century plant
(Agave americana).

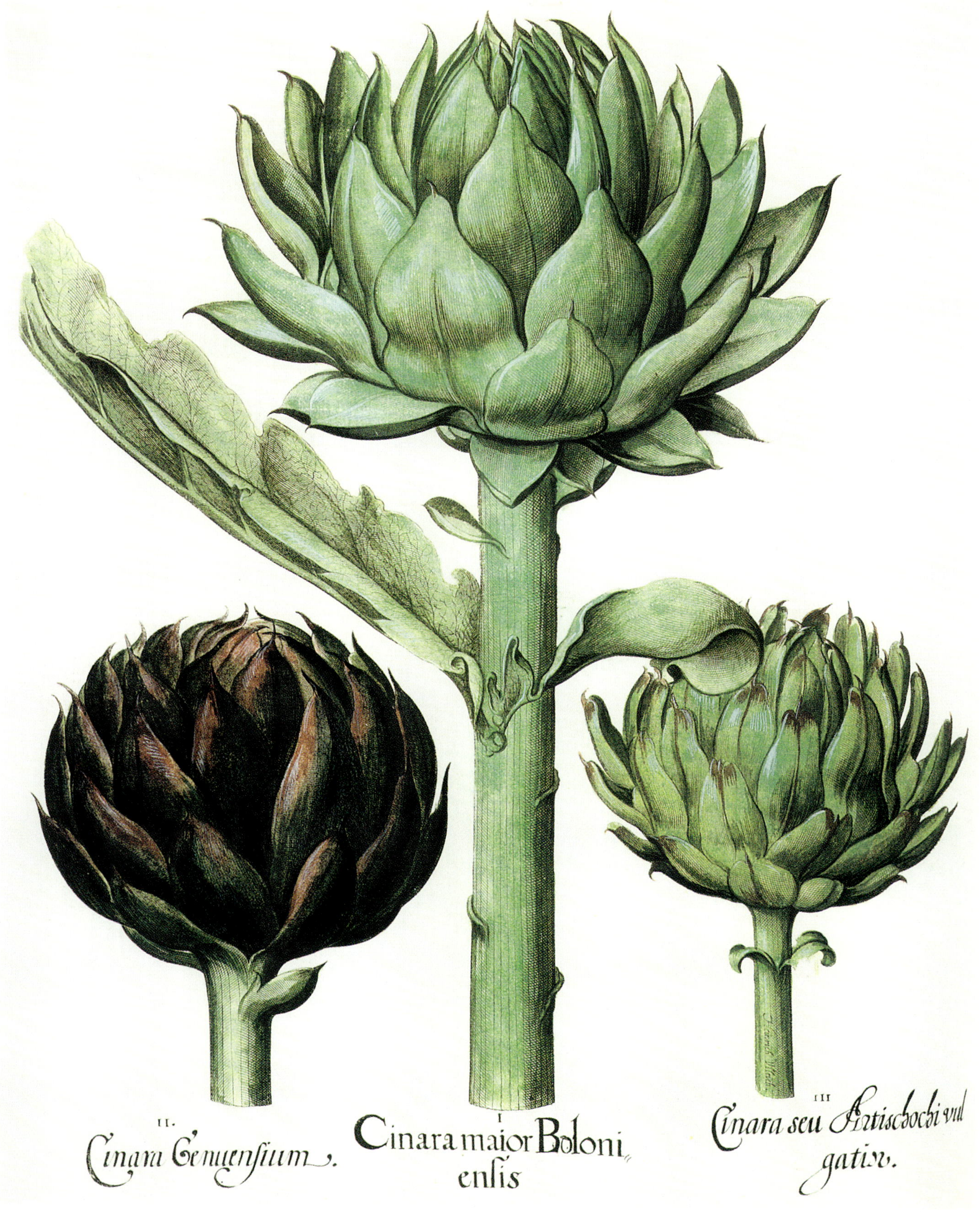

I, II, III. Globe artichoke
(Cynara scolymus).

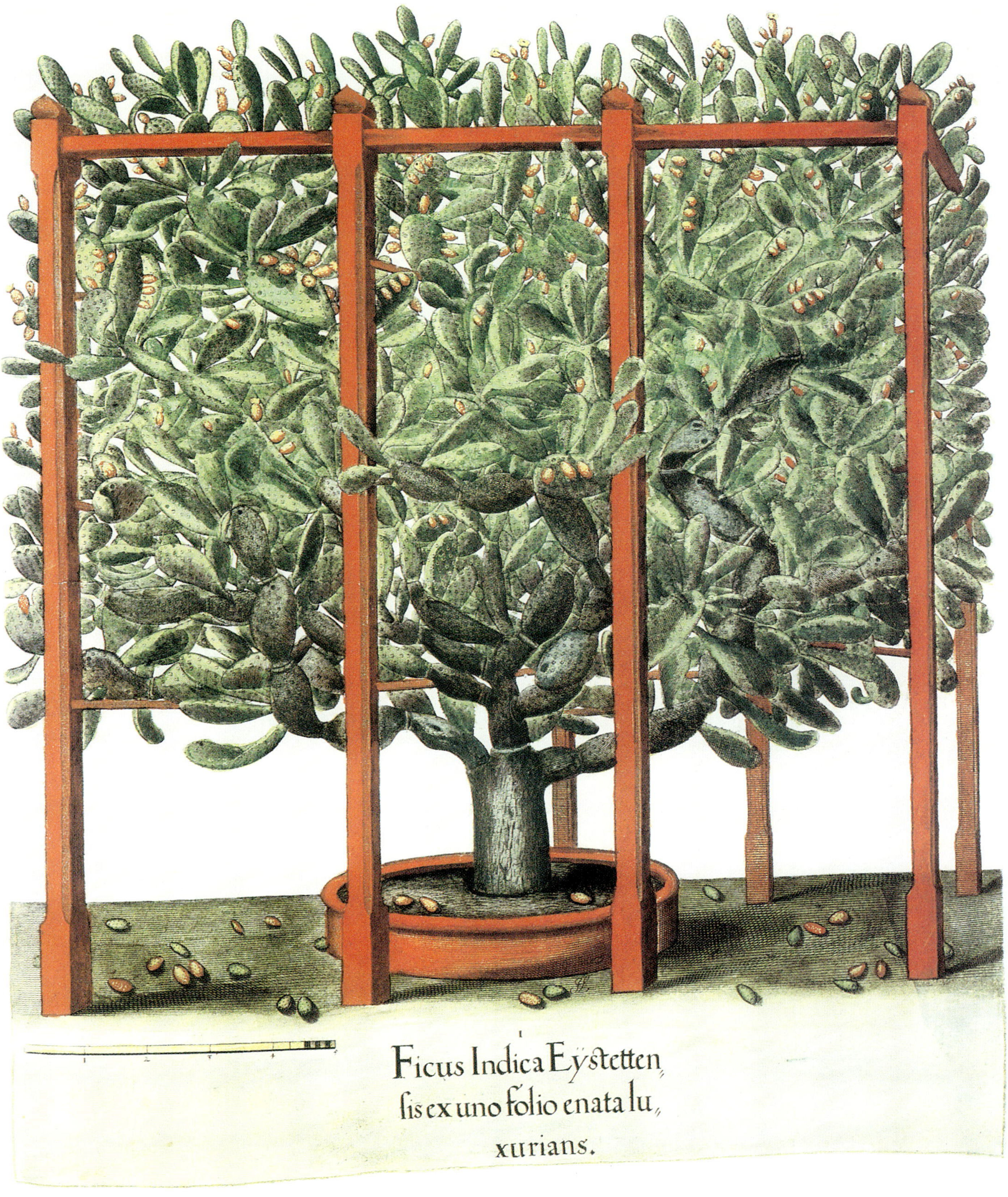

I. Indian fig
(Opuntia ficus-indica).

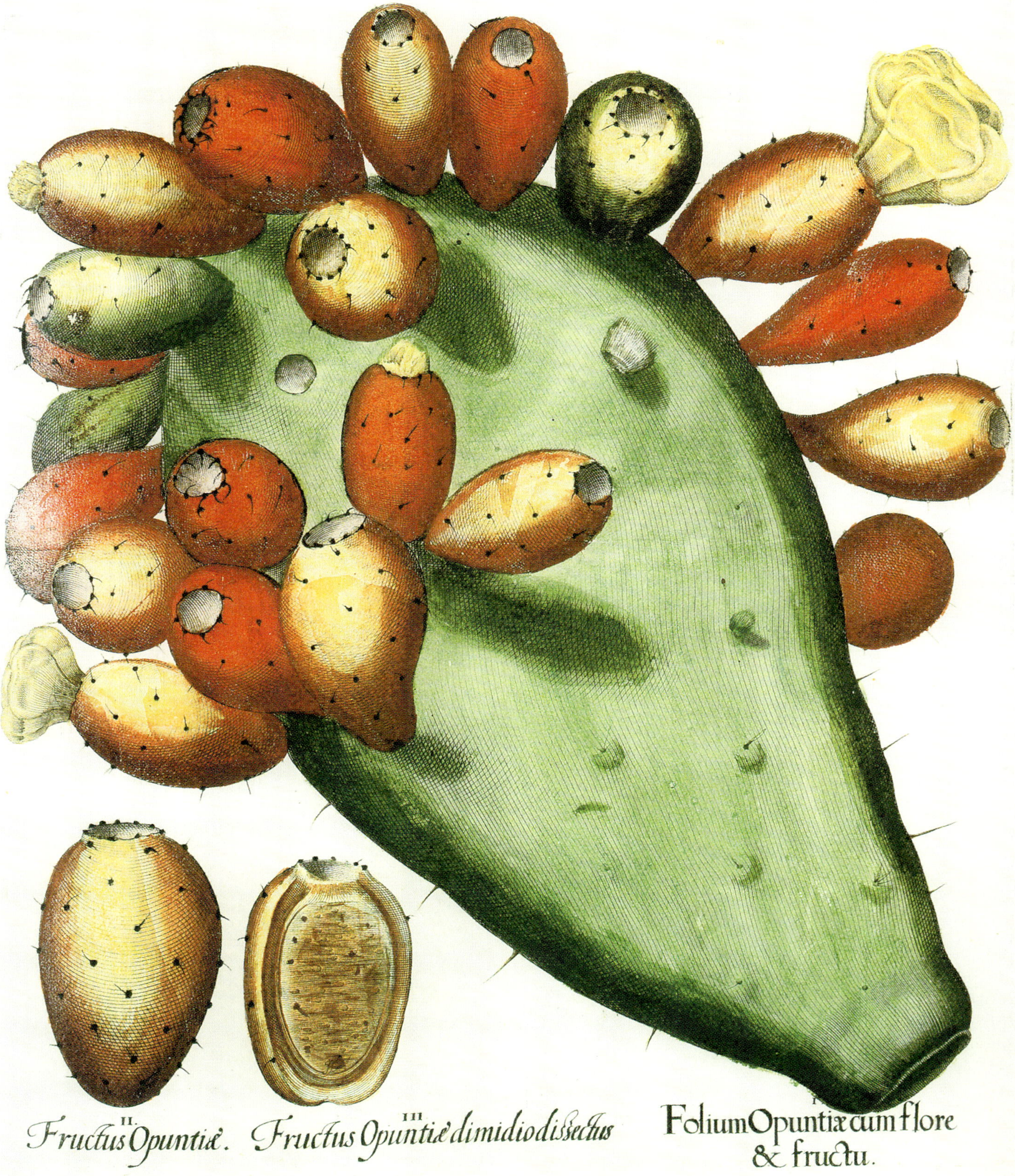

I, II, III. Indian fig, stem segment and fruits
(Opuntia ficus-indica).

Index

Acanthus mollis 43 / III
Aeonium arboreum 67 / I
Agave americana 69
Alcea rosea 40 / I, II
Amaranth 60 / I
Amaranthus paniculatus 60 / I
Anemone coronaria 3 / I
Anemone hortensis 3 / II, III
Anemone, blue 4 / IV, V
Anthemis cotula 27 / II
Apricot 26 / III
Artichoke, globe 70
Arum maculatum 62 / III
Arum, dragon 37 / I
Asarabacca 17 / II
Asarum europaeum 17 / II

Basil, common 51 / II, III
Bay, rose 23 / I, II
Bear's breeches 43 / III
Bellis spec. 48 / II, III
Bluebell 49 / I

Campanula medium 31 / I
Campanula rotundifolia 49 / I
Campion 31 / III
Campion, bladder 31 / II
Canterbury bells 31 / I
Capsicum spec. 57
Cardoon 45
Carnation, wild 52 / I
Carthamus tinctorius 44 / I
Castor oil plant 41 / I
Catchfly, Spanish 60 / III
Celandine, lesser 21 / II
Centaurium erythraea 14 / II, III
Centaury, common 14 / II, III
Century plant 69
Chamaemelum nobile 44 / II, III
Chamomile 44 / II
Cheiranthus cheiri 33 / I–III
Cherry plum 26 / II
Cherry, Jerusalem 29 / I
Cherry, winter 29 / I
Chili pepper 57
Citron 24 / II
Citrullus lanatus 53 / III
Citrus aurantium 24 / I
Citrus medica 24 / II
Citrus sinensis 24 / III
Cocoyam 65 / I
Colchicum spec. 67 / II, III
Colocasia esculenta 65 / I
Crepis tectorum 60 / II
Crocus 4 / II, III
Crocus sativus 36 / IV
Crocus spec. 36 / III
Crocus vernus 4 / II, III
Crocus, autumn 67 / II, III
Crocus, Dutch 36 / III
Crocus, saffron 36 / IV
Crown imperial 11; 12
Cucumis melo 53 / II
Cucurbita pepo 65 / III
Currant, Alpine 1 / I
Currant, black 1 / III
Currant, mountain 1 / I
Currant, red 1 / II, IV (IIII), V
Cyclamen hederifolium 66 / I
Cyclamen purpurascens 66 / II, III
Cynara cardunculus 45
Cynara scolymus 70

Dactylorhiza majalis 5 / III
Daffodil 6 / I, II
Daffodil, wild 6 / III
Daisy 48 / II, III
Datura metel 62 / I
Day lily 22 / I
Dianthus caryophyllus 52 / I
Digitalis purpurea 30 / I, II
Dracunculus vulgaris 37 / I

Eggplant 55

Fig, Indian 71; 72
Foxglove 30 / I, II
Fragaria spec. 18 / I–III

Fritillaria imperialis 11; 12

Gentian, chiltern 68 / III
Gentian, fringed 68 / II
Gentiana germanica 68 / III
Gentianella ciliata 68 / II
Gillyflower 32 / I–III
Gourd, white-flowered 65 / II
Groundsel 22 / II

Halimione portulacoides 62 / II
Hartwort 47 / I
Hawk's-beard 60 / II
Hawkweed, orange 30 / III
Helianthus annuus frontispiece
Helianthus x *multiflorus* 39
Hemerocallis fulva 22 / I
Hepatica nobilis 4 / IV, V
Hibiscus syriacus 27 / I; 28 / I
Hieracium aurantiacum 30 / III
Hollyhock 40 / I, II
Horn of plenty 62 / I
Houseleek 67 / I
Houseleek, Alpine 16 / II

Indian shot 58
Iris 19 / I; 38 / II, III
Iris florentina 19 / II
Iris germanica 20 / I–III
Iris pallida 19 / III
Iris spec. 19 / I
Iris xiphioides 38 / I
Iris xiphium 36 / I; 38 / II, III
Iris, Dalmatian 19 / III
Iris, English 38 / I
Iris, German 20 / I–III
Iris, Spanish 36 / I

Jasmine 43 / II
Jasmine, common 43 / I
Jasminum fruticans 43 / II
Jasminum grandiflorum 43 / I
Jasminum odoratissimum 43 / II
Jasminum officinale 43 / I

Lagenaria spec. 65 / II
Lavandula spec. 52 / II, III
Lavender 52 / II, III
Lilium bulbiferum 13 / I, II; 14 / I
Lilium chalcedonicum 35 / I, II; 36 / II
Lilium martagon 34
Lily, fire 13 / I, II; 14 / I
Linaria alpina 29 / II
Lords-and-ladies 62 / III
Lychnis chalcedonica 42 / I–III
Lycopersicon esculentum 54
Lycopersicon spec. 53 / I
Lycopodium clavatum 41 / II

Malcolmia maritima 27 / III
Mallow 27 / I; 28 / I
Mallow, rose 28 / I
Maltese cross 42 / I–III
Mandragora autumnalis 21 / I
Mandrake, autumn 21 / I
Marigold, African 50 / I; 51 / I
Marigold, French 50 / II, III
Marjoram 56 / II
Marvel of Peru 68 / I
Matthiola incana 32 / I–III
Mayweed 27 / II
Melon 53 / II
Mirabilis jalapa 68 / I
Morning glory 49 / II, III

Narcissus x *incomparabilis* 6 / I, II
Narcissus pseudonarcissus 6 / III
Nasturtium 48 / I
Nerium oleander 23 / I, II
Nicotiana latissima 61
Nightshade 56 / I

Ocimum basilicum 51 / II, III
Oleander 23 / I, II
Onion, sea 4 / I
Opuntia ficus-indica 71; 72
Orange, bitter 24 / I
Orange, common 24 / III
Orange, Seville 24 / I
Orchid, marsh 5 / III
Orchid, toothed 5 / II
Orchis tridentata 5 / II
Origanum majorana 56 / II
Orris root 19 / II

Paeonia officinalis 16 / I; 17 / I
Papaver rhoeas 46 / II
Papaver somniferum 46 / I; 47 / II, III

Peony, common 16 / I; 17 / I
Pharbitis nil 49 / III
Pharbitis purpurea 49 / II
Pine, ground 41 / II
Plantago major 25 / II, III
Plantain, common 25 / II, III
Pomegranate 25 / I; 26 / I
Poppy, corn 46 / II
Poppy, opium 46 / I; 47 / II, III
Potato 64 / I
Prunus armeniaca 26 / III
Prunus cerasifera 26 / II
Pumpkin 65 / III
Punica granatum 25 / I; 26 / I
Purslane, sea 62 / II

Ranunculus ficaria 21 / II
Ribes alpinum 1 / I
Ribes nigrum 1 / III
Ribes rubrum 1 / II, IV (IIII), V
Ricinus communis 41 / I
Rosa centifolia 15 / I
Rosa gallica officinalis 15 / III
Rosa gallica officinalis-versicolor 15 / II
Rosa hemisphaerica 15 / IV (IIII)
Rosa mundi 15 / II
Rose, apothecary's 15 / III
Rose, Provence 15 / I
Rose, sulphur 15 / IV (IIII)

Safflower 44 / I
Scilla autumnalis 35 / III
Scilla peruviana 5 / I
Sempervivum montanum 16 / II
Senecio doria 22 / II
Silene nutans 31 / III
Silene otites 60 / III
Silene vulgaris 31 / II
Solanum macrocarpon 56 / I
Solanum melongena 55
Solanum pseudocapsicum 29 / I
Solanum tuberosum 64 / I
Sorghum 63 / I, II
Sorghum bicolor 63 / I, II
Sowbread 66 / I–III
Squill of Peru 5 / I
Squill, autumn 35 / III
Stock, Virginia 27 / III
Strawberry 18 / I–III
Sunflower 39
Sunflower, common frontispiece

Tagetes erecta 50 / I; 51 / I
Tagetes patula 50 / II, III
Tampala 59
Thyme, broad-leaved 64 / II
Thyme, common 64 / III
Thymus pulegioides 64 / II
Thymus vulgaris 64 / III
Toadflax, Alpine 29 / II
Tobacco 61
Tomato 53 / I; 54
Tordylium apulum 47 / I
Tordylium officinale 47 / I
Tropaeolum minus 48 / I
Tulip 7–10
Tulipa spec. 7–10
Turk's-cap lily 34; 35 / I, II; 36 / II

Urginea maritima 4 / I

Viola odorata 2 / I–V
Violet, English 2 / I–V
Violet, sweet 2 / I–V

"Wallflower of Eichstätt" 33 / I
Wallflower 33 / I–III
Watermelon 53 / III
Windflower 3 / I–III